T0222400

Artificial Intelligence and Brain Research

Patrick Krauss

Artificial Intelligence and Brain Research

Neural Networks, Deep Learning and the Future of Cognition

 Springer

Patrick Krauss
University of Erlangen-Nuremberg
Erlangen, Germany

ISBN 978-3-662-68979-0   ISBN 978-3-662-68980-6   (eBook)
https://doi.org/10.1007/978-3-662-68980-6

Translation from the German language edition: "Künstliche Intelligenz und Hirnforschung" by Patrick Krauss, © Der/die Herausgeber bzw. der/die Autor(en), exklusiv lizenziert an Springer-Verlag GmbH, DE, ein Teil von Springer Nature 2023. Published by Springer Berlin Heidelberg. All Rights Reserved.

This book is a translation of the original German edition "Künstliche Intelligenz und Hirnforschung" by Patrick Krauss, published by Springer-Verlag GmbH, DE in 2023. The translation was done with the help of an artificial intelligence machine translation tool. A subsequent human revision was done primarily in terms of content, so that the book will read stylistically differently from a conventional translation. Springer Nature works continuously to further the development of tools for the production of books and on the related technologies to support the authors.

This Springer imprint is published by the registered company Springer-Verlag GmbH, DE, part of Springer Nature.
The registered company address is: Heidelberger Platz 3, 14197 Berlin, Germany

If disposing of this product, please recycle the paper.

For Sofie and Hannes

Preface

How does Artificial Intelligence work? How does the brain function? What are the similarities between natural and artificial intelligence, and what are the differences? Is the brain a computer? What are neural networks? What is Deep Learning? Should we attempt to recreate the brain to create real general Artificial Intelligence, and if so, how should we best proceed?

We are in an extremely exciting phase of cultural and technological development of humanity. Recently, Artificial Intelligence (AI) and Machine Learning have been making their way into more and more areas, such as medicine, science, education, finance, engineering, entertainment, and even art and music, and are becoming ubiquitous in twenty-first-century life. Particularly in the field of so-called Deep Learning, the progress is extraordinary in every respect, and deep artificial neural networks show impressive performance in a variety of applications such as processing, recognition, and generation of images or natural language. Especially in combination with a method called Reinforcement Learning, the networks are becoming increasingly powerful, for example when it comes to playing video games, or they even achieve superhuman abilities in complex board games like Go, when they are trained by playing millions of games against themselves.

Many of the algorithms, design principles, and concepts used in AI today, such as neural networks or the aforementioned reinforcement learning, have their origins in biology and psychology. Therefore, neuroscience lectures are becoming an integral part of courses such as computer science or artificial intelligence at more and more universities. But it is also worthwhile for brain researchers to engage with artificial intelligence, as it not only provides important tools for data evaluation, but also serves as a model for natural intelligence and has the potential to revolutionize our understanding of the brain.

When considering the goals of AI and neuroscience, it becomes apparent that they are complementary to each other. The goal of AI is to achieve cognition and behavior at a human level, and the goal of neuroscience is to understand human cognition and behavior. One could therefore say that artificial intelligence and brain research are two sides of the same coin. The convergence of both research fields promises profound synergies, and it is already certain that the insights gained from this will shape our future in a sustainable way.

In recent years, I have given many lectures on these and related topics. From the subsequent discussions and numerous follow-up questions, I learned that the deep connection between AI and brain research is immediately apparent to most, but was not really conscious before. Although this is gradually beginning to change, most people associate AI exclusively with degree programs such as computer science or data science, and less so with cognitive science or computational neuroscience, even though these branches of science can contribute a lot to basic research in AI. Conversely, artificial intelligence has become indispensable in modern brain research. To understand how the human brain works, research teams are increasingly using models based on artificial intelligence methods, gaining not only neuroscientific insights, but also learning something about artificial intelligence.

There are already many excellent textbooks and non-fiction books in which the various disciplines are each presented in isolation. However, an integrated presentation of AI and brain research has not yet existed. With this book, I want to close this gap. Based on exciting and current research results, the basic ideas and concepts, open questions, and future developments at the intersection of AI and brain research are clearly presented. You will learn how the human brain is structured, what fundamental mechanisms perception, thinking, and action are based on, how AI works, and what is behind the spectacular achievements of AlphaGo, ChatGPT, and Co. Please note that I am not aiming for a comprehensive introduction to AI or brain research. You should only be equipped with what I consider to be the theoretical minimum, so that you can understand the challenges, unsolved problems, and ultimately the integration of both disciplines.

The book is divided into four parts, some of which build on each other, but can also be read independently of each other. So there are different ways to approach the content of this book. Of course, I would prefer if you read the book as a whole, preferably twice: once to get an overview, and a second time to delve into the details. If you want to get an overview of how the brain works, then start with Part I. However, if you are more interested in getting an overview of the state of research in Artificial Intelligence, then I recommend you start with Part II. The open questions and challenges of both disciplines are presented in Part III. If you are already familiar with the basics and open questions of AI and brain research and are primarily interested in the integration of both research branches, then read Part IV.

I have tried to clarify complex issues through illustrative diagrams wherever possible. My children have actively supported me in creating these diagrams. English quotes have been translated by me, unless otherwise noted. Colleagues, friends, and relatives have greatly helped in correcting errors and improving the clarity and readability of the text. I would like to thank Konstantin Tziridis, Claus Metzner, Holger Schulze, Nathaniel Melling, Tobias Olschewski, Peter Krauß, and Katrin Krauß for this.

My special thanks go to Sarah Koch, Ramkumar Padmanaban, and Ken Kissinger from Springer Publishing, who have supported me in the realization of this book project.

My research work has been and continues to be supported by the German Research Foundation. I am grateful to those in charge. Without the inspiring working atmosphere at the Friedrich-Alexander University Erlangen-Nuremberg and the University Hospital Erlangen, many of my ideas and research projects would not have been possible. My special thanks go to Holger Schulze, Andreas Maier, and Thomas Herbst for their support, as well as Claus Metzner and Achim Schilling for the countless inspiring conversations. I sincerely thank my father for the many discussions on the various topics of this book. My greatest thanks go to my wife, who has always supported everything over the years and continues to do so. What I owe her, I cannot put into words. I dedicate this book to my children.

Großenseebach   Patrick Krauss
in April 2023

Contents

1

Introduction

An elephant is like a fan!
The fifth blind man

ChatGPT Passes the Turing Test

In the field of Artificial Intelligence (AI), there have been a number of spectacular breakthroughs in the last approximately 10–15 years—from *AlphaGo* to *DALL-E 2*to *ChatGPT* –, which until recently were completely unthinkable.

The most recent event in this series is certainly also the most spectacular: It is already clear that November 30, 2022 will go down in history. On this day, the company OpenAI made the Artificial Intelligence *ChatGPT* freely accessible to the public. This so-called Large Language Model can generate any type of text in seconds, answers questions on any topic, gives interviews and conducts conversations, remembering the course of which and thus responding adequately even in longer conversations. Since then, millions of people have been able to convince themselves daily of the amazing capabilities of this system. The responses and texts generated by *ChatGPT* are indistinguishable from those produced by humans. *ChatGPT* thus becomes the first in the history of Artificial Intelligence to pass the Turing Test, a procedure devised to determine whether a machine has the ability to think (Turing, 1950). An artificial system that passes the Turing Test has

© The Author(s), under exclusive license to Springer-Verlag GmbH, DE, part of Springer Nature 2024
P. Krauss, *Artificial Intelligence and Brain Research*,
https://doi.org/10.1007/978-3-662-68980-6_1

long been considered the Holy Grail of research in the field of Artificial Intelligence. Even though passing the Turing Test does not necessarily mean that *ChatGPT* actually thinks, you should still remember November 30, 2022 well. It not only represents the most important milestone in the history of Artificial Intelligence to date, but its significance is certainly comparable to the invention of the loom, the steam engine, the automobile, the telephone, the internet, and the smartphone, which often only turned out to be game-changers and decisive turning points in development in retrospect.

The Next Affront

In addition to the much-discussed consequences that *ChatGPT* and similar AI systems will have on almost all levels of our societal life, the astonishing achievements of these new systems also challenge our explanations of what fundamental concepts such as cognition, intelligence andconsciousness mean at all. In particular, the influence that this new type of AI will have on our understanding of the human brain is already immense and its impacts are not yet fully foreseeable.

Some are already talking about the next great affront to humanity. These are fundamental events or insights that have profoundly shaken man's self-understanding and his relationship to the world throughout history.

The Copernican affront, named after the astronomer Nicolaus Copernicus, refers to the discovery that the Earth is not the center of the universe, but revolves around the sun. This realization in the sixteenth century fundamentally changed the world view and led to a loss of self-centeredness and self-assurance. With the discovery of thousands of exoplanets, i.e., planets outside our solar system, in recent decades, this affront has even been intensified. This has shown that planetary systems are very common in our galaxy and that there may even be many planets that orbit in the habitable zone around their stars and thus represent possible places for life.

Another insight that affected man's self-understanding was the Darwinian affront. Charles Darwin's theory of evolution in the nineteenth century showed that man is not a species created by God, but has evolved like all other species through evolution. This discovery questioned man's self-understanding as a unique species separated from the rest of nature.

Another affront, which Sigmund Freud modestly named after the theory he developed, the psychoanalytic affront, refers to the discovery that human behavior and thinking are not always consciously and rationally controlled, but are also influenced by unconscious and irrational drives. This realization shook man's confidence in his ability for self-control and rationality. The Libet experiments, which even question the existence of free will, further intensified the impact of this affront.

AI can be considered as a newly added fourth major affront to human self-understanding. Until now, our highly developed language was considered the decisive distinguishing feature between humans and other species. However, the development of large language models like *ChatGPT* has shown that machines are in principle capable of dealing with natural language in a similar way to humans. This fact challenges our concept of uniqueness and incomparability as a species again and forces us to at least partially rethink our definition of being human.

This "AI affront" affects not only our linguistic abilities, but our cognitive abilities in general. AI systems are already capable of solving complex problems, recognizing patterns, and achieving human-like or even superhuman performance in certain areas (Mnih et al., 2015; Silver et al., 2016, 2017a, b; Schrittwieser et al., 2020; Perolat et al., 2022). This forces us to reinterpret human intelligence and creativity, where we have to ask ourselves what role humans play in a world where machines can take over many of our previous tasks. It also forces us to think about the ethical, social, and philosophical questions that arise from the introduction of AI into our lives. For example, the question arises as to how we should deal with the responsibility for decisions made by AI systems, and what limits we should set on the use of AI to ensure that it serves the good of humanity (Anderson & Anderson, 2011; Goodall, 2014; Vinuesa et al., 2020).

Less than half a year after the publication of *ChatGPT*, its successor *GPT-4* was released in March 2023, which significantly surpasses the performance of its predecessor. This prompted some of the most influential thinkers in this field to call for a temporary pause in the further development of AI systems, which are even more powerful than GPT-4, in a widely noticed open letter[1] to prevent a potentially impending loss of control.

Artificial Intelligence and Brain Research

The remarkable achievements of *ChatGPT* and *GPT-4* also have direct implications for our understanding of the human brain and how it functions. They therefore not only challenge brain research, but even have the

---

[1] https://futureoflife.org/open-letter/pause-giant-ai-experiments/

potential to revolutionize it. Indeed, AI and brain research have always been closely intertwined in their history. The so-called cognitive revolution in the middle of the last century can also be seen as the birth of research in the field of AI, where it developed as an integral part of the newly emerged research agenda of cognitive sciences as an independent discipline. In fact, AI research was never just about developing systems to take over tedious work. From the beginning, it was also about developing and testing theories of natural intelligence. As we will see, some astonishing parallels between AI systems and brains have been uncovered recently. Therefore, AI plays an increasingly important role in brain research, not only as a pure tool for data analysis, but especially as a model for the function of the brain.

Conversely, neuroscience has also played a key role in the history of artificial intelligence and has repeatedly inspired the development of new AI methods. The transfer of design and processing principles from biology to computer science has the potential to provide new solutions for current challenges in the field of AI. Here too, brain research not only plays the role of providing the brain as a model for new AI systems. Rather, a variety of methods for deciphering the representation and calculation principles of natural intelligence have been developed in neuroscience, which can now in turn be used as a tool for understanding artificial intelligence and thus contribute to solving the so-called black box problem. An endeavor occasionally referred to as Neuroscience 2.0. It is becoming apparent that both disciplines will increasingly merge in the future (Marblestone et al., 2016; Kriegeskorte & Douglas, 2018; Rahwan et al., 2019; Zador et al., 2023).

Too Blind to See the Elephant

The realization that different disciplines must work together to understand something as complex as human-level cognition is of course not new and is vividly illustrated in the well-known metaphor of the six blind men and the elephant (Friedenberg et al., 2021):

> Once upon a time, there were six blind scientists who had never seen an elephant and wanted to research what an elephant is and what it looks like. Each examined a different part of the body and accordingly came to a different conclusion.
>
> The first blind approached the elephant and touched its side. "Ah, an elephant is like a wall," she said.

The second blind touched the elephant's tusk and exclaimed: "No, an elephant is like a spear!"

The third blind touched the elephant's trunk and said: "You are both wrong! An elephant is like a snake!"

The fourth blind touched a leg of the elephant and said: "You are all wrong. An elephant is like a tree trunk."

The fifth blind touched the elephant's ear and said: "None of you know what you're talking about. An elephant is like a fan."

Finally, the sixth blind approached the elephant and touched its tail: "You are all wrong," he said. "An elephant is like a rope."

If the six scientists had combined their findings, they would have come much closer to the true nature of the elephant. In this story, the elephant represents the human mind, and the six blind people represent the various scientific disciplines that try to understand its functioning from different perspectives (Fig. 1.1). The punchline of the story is that while each individual's perspective is valuable, a comprehensive understanding of cognition

Fig. 1.1 The Blind Men and the Elephant. Each examines a different part of the body and accordingly comes to a different conclusion. The elephant represents the mind and brain, and the six blind represent different sciences. The perspective of each individual discipline is valuable, but a comprehensive understanding can only be achieved through collaboration and interdisciplinary exchange

can only be achieved when the different sciences work together and exchange ideas.

This is the founding idea of cognitive science, which began in the 1950s as an intellectual movement referred to as the cognitive revolution (Sperry, 1993; Miller, 2003). During this time, there were significant changes in the way psychologists and linguists worked and new disciplines such as computer science and neuroscience emerged. The cognitive revolution was driven by a number of factors, including the rapid development of personal computers and new imaging techniques for brain research. These technological advances allowed researchers to better understand how the brain works and how information is processed, stored, and retrieved. As a result of these developments, an interdisciplinary field emerged in the 1960s that brought together researchers from a wide range of disciplines. This field went by various names, including information processing psychology, cognition research, and indeed cognitive science.

The cognitive revolution marked a significant turning point in the history of psychology and related disciplines. It fundamentally changed the way researchers approach questions of human cognition and behavior, paving the way for numerous breakthroughs in areas such as artificial intelligence, cognitive psychology, and neuroscience.

Today, cognitive science is understood as an interdisciplinary scientific endeavor to explore the different aspects of cognition. These include language, perception, memory, attention, logical thinking, intelligence, behavior and emotions. The focus is primarily on the way natural or artificial systems represent, process, and transform information (Bermúdez, 2014; Friedenberg et al., 2021).

The key questions are: How does the human mind work? How does cognition work? How is cognition implemented in the brain? And how can cognition be implemented in machines?

Thus, cognitive science addresses some of the most difficult scientific problems, as the brain is incredibly difficult to observe, measure, and manipulate. Many scientists even consider the brain to be the most complex system in the known universe.

The disciplines involved in cognitive science today include linguistics, psychology, philosophy, computer science, artificial intelligence, neuroscience, biology, anthropology, and physics (Bermúdez, 2014). For a time, cognitive science fell somewhat out of fashion, particularly the idea of integrative collaboration between different disciplines was somewhat forgotten. Specifically, AI and neuroscience developed independently and thus also away from each other. Fortunately, the idea that AI and brain research are complementary and can benefit greatly from each other is currently experiencing a real renaissance, with the term "cognitive science" apparently being interpreted differently in some communities today or considered too old-fashioned, which is why terms like *Cognitive Computational Neuroscience* (Kriegeskorte & Douglas, 2018) or *NeuroAI* (Zador et al., 2023) have been suggested instead.

The legacy of the cognitive revolution is evident in the many innovative and interdisciplinary approaches that continue to shape our understanding of the human mind and its functioning. Whether through state-of-the-art brain imaging techniques, sophisticated computer models, or new theoretical frameworks—researchers are constantly pushing the boundaries of what we know about the human brain and its complex processes.

Brain-Computer Analogy

Many researchers believe that computer models of the mind can help us understand how the brain processes information, and that they can lead to the development of more intelligent machines. This assumption is based on the brain-computer analogy (Von Neumann & Kurzweil, 2012). It is assumed that mental processes such as perception, memory, and logical thinking involve the manipulation of mental representations that correspond to the symbols and data structures used in computer programs (Fig. 1.2). Like a computer, the brain is capable of receiving, storing, processing, and outputting information.[2]

However, this analogy does not mean that the brain is actually a computer, but that it performs similar functions. By considering the brain as a computer, one can abstract from biological details and focus on the way it processes information to develop mathematical models for learning, memory, and other cognitive functions.

The brain-computer analogy is based on two central assumptions that underlie cognitive science. These are computationalism and functionalism.

---

[2]A fundamental difference is that a computer processes information with different components than those with which it stores the information. In the brain, both are done by the—sometimes same—neurons.

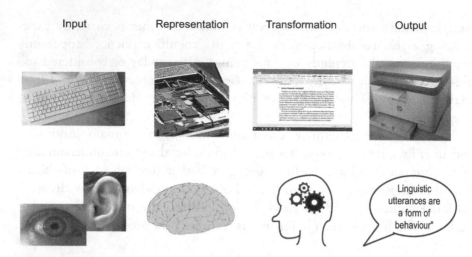

Fig. 1.2 Brain-Computer Analogy. Information processing includes the input, representation, transformation, and output of information. For a computer, the input may come from the keyboard, for a biological organism from the sensory organs. This input must then be represented: by storing it on a hard drive or in the computer's RAM, or in the brain as momentary neuronal activity in short-term memory or in long-term memory in the interconnection of neurons. Then a transformation or processing takes place, i.e., mental processes or algorithms must act on the stored information and change it to generate new information. For a computer, this could be text processing, for humans, for example, logical reasoning. Finally, the result of information processing is output. The output can be, for example, via a printer for a computer. In living beings, the output corresponds to observable behavior or, as a special case of behavior, to human linguistic utterances

Computationalism

In computationalism, it is assumed that cognition is synonymous with information processing, i.e., that mental processes can be understood as calculations and that the brain is essentially an information processing system (Dietrich, 1994; Shapiro, 1995; Piccinini, 2004, 2009). Like any such system, the brain must therefore represent information and then transform these represented information, i.e., there must be mental representations of information and there must be mental processes that can act on these representations and manipulate them. Computationalism has greatly influenced the way cognitive scientists and researchers in the field of artificial intelligence think about intelligence and cognition.

However, there is also criticism of this view, as evidenced by numerous ongoing debates in philosophy and cognitive science. Some critics argue, for example, that the computer model of the mind is too simple and cannot fully capture the complexity and richness of human cognition. Others argue that it is unclear whether mental processes can really be understood as calculations or whether they fundamentally differ from the way processes occur in computers.

Functionalism

Is cognition only possible in a (human) brain? Functionalism clearly answers this question with a no. Accordingly, mental states and processes are defined exclusively by their functions or their relationship to behavior, not by their physical or biochemical properties (Shoemaker, 1981; Jackson & Pettit, 1988; Piccinini, 2004). What does this mean in concrete terms?

Please imagine a car in your mind's eye. And now remember the last situation in which you ate chocolate, and try to recall the taste as accurately as possible. Did you succeed? I assume you did. As I write these lines, I have brought to mind the same two mental states *"seeing a car"* and *"tasting chocolate"*. Obviously, each of us can activate the corresponding mental representations in our brains, even though you, I, and every other reader of these lines have completely different brains. All human brains are of course similar in their basic structure. But they are certainly not identical down to the smallest detail, especially not in the exact wiring of the neurons, if only because every person has had completely different, individual experiences in their life, which affect the wiring pattern of the brain. In computer science terminology, one would say that each person has a different, individual hardware. Yet we can all bring to mind the same mental state.

While in the previous example the systems were somehow very similar— they were always human brains—the following example may illustrate how much the different physical implementations of the same algorithm can differ from each other. Consider the addition of two numbers. The representation of these numbers, as well as the associated process or algorithm to add them, can be implemented in your brain when you "calculate in your head", or for example also in a laptop with spreadsheet program, a slide rule, a calculator or a calculator app on your smartphone. Each time, the same numbers are represented and added, while the information processing systems are completely different. This is the concept of *multiple realizability*.

Accordingly, the same mental state or process can in principle be realized by completely different natural or artificial systems. Put simply, this means that cognition and presumably also consciousness can in principle be implemented in any physical system capable of supporting the required

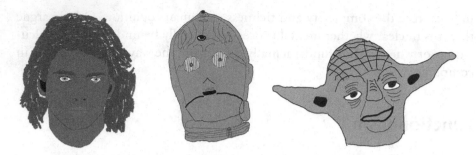

Fig. 1.3 Functionalism. Human-level cognition is not limited to a human brain, but could in principle also be implemented in any other system that supports the required calculations, such as correspondingly advanced robots or aliens

calculations. If many different human brains are already capable of this, why should this ability be limited to humans or biological systems? From the perspective of functionalism, it is therefore quite possible that the ability for human-like cognition can also be implemented in correspondingly highly developed machines or alien brains (Fig. 1.3).

Conclusion

In recent years, spectacular advances in artificial intelligence have turned our understanding of cognition, intelligence, and consciousness upside down and will have profound impacts on society and our understanding of the human brain. Cognitive science is the key to a deeper understanding of brain and mind, and computer models of the mind can help us understand how the brain processes information and contribute to the development of smarter machines. These models are based on the central assumptions of computationalism and functionalism, which emphasize the equivalence of cognition and information processing as well as the independence of cognitive processes from their physical implementation.

The advances in artificial intelligence have also led to the fields of neuroscience and computer science becoming increasingly intertwined. The transfer of construction and processing principles from biology to computer science promises new solutions for current challenges in artificial intelligence. Conversely, the close collaboration of these disciplines will become increasingly important in the future to understand complex systems like the human brain.

The recent advances in artificial intelligence and their applications have opened the door in neuroscience to new insights and technologies far beyond what was previously possible. We are only at the beginning of a new era of research and innovation, and it remains to be seen what fascinating discoveries and developments await us in the future.

References

Anderson, M., & Anderson, S. L. (Hrsg.). (2011). *Machine ethics.* Cambridge University Press.

Bermúdez, J. L. (2014). *Cognitive science: An introduction to the science of the mind.* Cambridge University Press.

Dietrich, E. (1994). *Computationalism. In thinking computers and virtual persons* (pp. 109–136). Academic.

Friedenberg, J., Silverman, G., & Spivey, M. J. (2021). *Cognitive science: An introduction to the study of mind.* Sage.

Goodall, N. J. (2014). Machine Ethics and Automated Vehicles. In G. Meyer & S. Beiker (Eds.), *Road vehicle automation. Lecture notes in mobility.* Springer. https://doi.org/10.1007/978-3-319-05990-7_9.

Jackson, F., & Pettit, P. (1988). Functionalism and broad content. *Mind,97*(387), 381–400.

Kriegeskorte, N., & Douglas, P. K. (2018). Cognitive computational neuroscience. *Nature neuroscience,21*(9), 1148–1160.

Marblestone, A. H., Wayne, G., & Kording, K. P. (2016). Toward an integration of deep learning and neuroscience. *Frontiers in computational neuroscience,10,* 94.

Miller, G. A. (2003). The cognitive revolution: A historical perspective. *Trends in cognitive sciences,7*(3), 141–144.

Mnih, V., Kavukcuoglu, K., Silver, D., Rusu, A. A., Veness, J., Bellemare, M. G., … & Hassabis, D. (2015). Human-level control through deep reinforcement learning. *Nature,518*(7540), 529–533.

Perolat, J., De Vylder, B., Hennes, D., Tarassov, E., Strub, F., de Boer, V., … & Tuyls, K. (2022). Mastering the game of Stratego with model-free multiagent reinforcement learning. *Science,378*(6623), 990–996.

Piccinini, G. (2004). Functionalism, computationalism, and mental contents. *Canadian Journal of Philosophy,34*(3), 375–410.

Piccinini, G. (2009). Computationalism in the philosophy of mind. *Philosophy Compass,4*(3), 515–532.

Rahwan, I., Cebrian, M., Obradovich, N., et al. (2019). Machine behaviour. *Nature,568,* 477–486.

Schrittwieser, J., Antonoglou, I., Hubert, T., Simonyan, K., Sifre, L., Schmitt, S., … & Silver, D. (2020). Mastering Atari, Go, Chess and Shogi by planning with a learned model. *Nature,588*(7839), 604–609.

Shapiro, S. C. (1995). Computationalism. *Minds and Machines,5*, 517–524.

Shoemaker, S. (1981). Some varieties of functionalism. *Philosophical topics,12*(1), 93–119.

Silver, D., Huang, A., Maddison, C. J., Guez, A., Sifre, L., Van Den Driessche, G., … & Hassabis, D. (2016). Mastering the game of Go with deep neural networks and tree search. *Nature,529*(7587), 484–489.

Silver, D., Schrittwieser, J., Simonyan, K., Antonoglou, I., Huang, A., Guez, A., … & Hassabis, D. (2017a). Mastering the game of Go without human knowledge. *Nature,550*(7676), 354–359.

Silver, D., Hubert, T., Schrittwieser, J., Antonoglou, I., Lai, M., Guez, A., … & Hassabis, D. (2017b). *Mastering Chess and Shogi by self-play with a general reinforcement learning algorithm.* arXiv preprint arXiv:1712.01815.

Sperry, R. W. (1993). The impact and promise of the cognitive revolution. *American Psychologist,48*(8), 878.

Turing, A. M. (1950). Computing machinery and intelligence. *Mind,59*(236), 433–460.

Vinuesa, R., Azizpour, H., Leite, I., Balaam, M., Dignum, V., Domisch, S., … & Fuso Nerini, F. (2020). The role of artificial intelligence in achieving the sustainable development goals. *Nature Communications,11*(1), 233.

Von Neumann, J., & Kurzweil, R. (2012). *The computer and the brain.* Yale University Press.

Zador, A., Escola, S., Richards, B., Ölveczky, B., Bengio, Y., Boahen, K., … & Tsao, D. (2023). Catalyzing next-generation artificial intelligence through NeuroAI. *Nature Communications,14*(1), 1597.

Part I

Brain Research

In the first part of the book, the aim is to familiarize you with the most important aspects of the structure and function of the brain. In doing so, a detailed and systematic description of many molecular biological, physiological, and anatomical details is deliberately avoided. The presentation also makes no claim to completeness. Interested readers may deepen their knowledge with one of the many excellent textbooks available on psychology and neuroscience. Rather, these first chapters are intended to convey the basics necessary from the author's point of view, on the basis of which we want to show the numerous cross-connections to Artificial Intelligence in later chapters.

2

The Most Complex System in the Universe

There is always a bigger fish.

Qui-Gon Jinn

The Brain in Numbers

The human brain consists of approximately 86 billion nerve cells, known as neurons (Herculano-Houzel, 2009). These are the fundamental processing units responsible for the reception, processing, and transmission of information throughout the body. The neurons are connected via so-called synapses and form a gigantic neural network. On average, each neuron receives its input from about 10,000 other neurons and sends, also on average, its output to about 10,000 subsequent neurons (Kandel et al., 2000; Herculano-Houzel, 2009). The actual number of connections per neuron can vary significantly, over several orders of magnitude, which is why we also speak of a broad distribution of connections per neuron. Some neurons, such as those in the spinal cord, are only connected to a single other neuron, while others, for example in the cerebellum, can be connected to up to a million other neurons.

Based on the total number of neurons and the average number of connections per neuron, the total number of synaptic connections in the brain can be roughly estimated at one quadrillion. This is a number with 15 zeros and can also be written as 10^{15}. In recent years, we have become somewhat

P. Krauss, *Artificial Intelligence and Brain Research*, https://doi.org/10.1007/978-3-662-68980-6_2

accustomed to amounts beyond a thousand billion, i.e., in the trillion range, in the context of politics and economics. The approximate number of synapses in the brain is a thousand times larger!

This may sound like a lot, but let's consider how many synapses would theoretically be possible in the brain. Each of the approx. 10^{11} neurons could in principle be connected to every other, with the information between any two neurons potentially running in two directions: either from neuron A to neuron B or vice versa. Additionally, each neuron can indeed be connected to itself. These special types of connections are called autapses. Purely combinatorially, this results in 10^{11} times 10^{11}, or 10^{22}, as the possible number of synapses. A comparison with the number of actually existing synapses shows that only about one in 10 million theoretically possible connections is actually realized. The network that the neurons form in the brain is therefore anything but dense *(dense),* but on the contrary extremely sparse *(sparse)* (Hagmann, 2008).

How Many Different Brains Can There Be?

In reference to the genome, which refers to the entirety of all genes of an organism, the connectome is the entirety of all connections in the nervous system of a living being (Sporns et al., 2005). To answer the question of how many different brains there can be, one must estimate how many different connectomes are combinatorially possible. At this point, it should be noted that not every theoretically possible connectome must result in a functioning viable nervous system. It turns out that it is quite complicated to calculate the exact number, which is why we want to settle here with an estimate for the lower limit of the actual number based on some simplifications. Let's assume for simplicity that each of the 10^{22} theoretically possible connections can either be present or not present. So we assume binary connections, with one representing an existing and zero a non-existing connection. As we will see later, reality is even more complicated. But even under this strong simplification, an absurdly high number of $2^{10^{22}}$ (read "two to the power of ten to the power of 22") results. This corresponds to a number with a trillion zeros. Of course, not every one of these connectomes leads to a powerful and viable nervous system, so the realizable number should only correspond to a tiny subset of all theoretically possible connectomes. On the other hand, the synaptic weights are not binary numbers, but can take any continuous value, which significantly increases the number of possibilities again.

How Many Different Mental States are Possible?

Let's now consider how many different brain or mental states there can be. Why is this important, you ask? Well, one of the central assumptions of modern cognitive and neuroscience is that every mental state (mental state) is based on a neuronal state (brain state) or a whole sequence of brain states. In other words, everything we can think, feel and experience has a specific neuronal correlate, i.e., an activation or even a temporal sequence of activations of neurons in our brain. This is also referred to as spatiotemporal activation patterns.

In this estimation, we again assume simplification. At any given moment, a specific neuron can either be active, i.e., send an action potential, which corresponds to one, or it does not do this and is thus inactive, which corresponds to zero. So here too, we end up with the binary system. As a further simplification, we divide the continuous flow of time into the smallest meaningful units. The typical duration of an action potential is one millisecond, i.e., one thousandth of a second. This is the characteristic time scale of the brain. Therefore, in each millisecond, each of the 10^{11} neurons in the brain can either be active or inactive. This results in a purely combinatorial $2^{10^{11}}$ different activation patterns per millisecond. Of course, we do not have a new thought, perception or feeling impression every millisecond or a thousand times per second. It is assumed that what we experience as now, the present or the moment, lasts about three seconds (Pöppel, 1997), which corresponds to 3000 milliseconds. Each mental state would therefore correspond to a sequence of 3000 different activation patterns within our simplifications. Since there can be $(2^{10^{11}})^{3000}$ different such sequences, our estimate results in just as many different brain states.

Conclusion

If you feel like the author, then you can't imagine anything meaningful under these orders of magnitude, as these numbers are essentially approaching infinity. While we were able to rewrite the number of possible brains or connectomes with a number with a trillion zeros, the number of mental states can no longer be put into words, as we would already have to use exponential notation for the number of digits of this number.

We can try to at least somewhat classify these absurd orders of magnitude. Apart from infinity, what is the largest number that still has a meaningful meaning in the natural sciences? It is 10^{82}. That's about how many atoms there are in the observable universe (Eddington, 1931). Or in other words: there are significantly fewer of the smallest in the largest than there are possible brains and mental states.

We can rightly claim that the human brain is probably the most complex system in the universe—that we know of. Of course, it is by no means excluded that there are far more complex systems somewhere in the vastness of the universe, such as the natural or possibly also artificial information processing systems of a highly developed species or artificial intelligence.

References

Eddington, A. S. (1931). Preliminary note on the masses of the electron, the proton, and the universe. In *Mathematical Proceedings of the Cambridge Philosophical Society* (Vol. 27, No. 1, pp. 15–19). Cambridge University Press.

Hagmann, P., Cammoun, L., Gigandet, X., Meuli, R., Honey, C. J., Van Wedeen, J., et al. (2008). Mapping the structural core of human cerebral cortex. *PLoS Biology, 6*(7), 1479–1493.

Herculano-Houzel, S. (2009). The human brain in numbers: A linearly scaled-up primate brain. *Frontiers in Human Neuroscience, 3*, 31.

Kandel, E. R., Schwartz, J. H., Jessell, T. M., Siegelbaum, S., Hudspeth, A. J., & Mack, S. (Eds.). (2000). *Principles of neural science* (Vol. 4, pp. 1227–1246). McGraw-Hill.

Sporns, O., Tononi, G., & Kötter, R. (2005). The human connectome: A structural description of the human brain. *PLoS Computational Biology, 1*(4), 0245–0251.

Pöppel, E. (1997). A hierarchical model of temporal perception. *Trends in Cognitive Sciences, 1*(2), 56–61.

3

Building Blocks of the Nervous System

There is not the slightest reason to doubt that brains are nothing more than machines with a huge number of parts, functioning in perfect accordance with the laws of physics.

Marvin Minsky

Neurons and Synapses

Neurons (nerve cells) are the fundamental processing units of every known nervous system (Kandel et al., 2000). The basic structure and function are the same in all neurons (Fig. 3.1). Like most other cells, they consist of a cell body with a nucleus, which contains the genetic information. What makes them unique, however, are the dendrites. These are many small, highly branched, antenna-like protrusions on the cell surface, through which the nerve cell receives signals from other nerve cells. Another characteristic is the axon, which represents the output channel of each neuron, through which it transmits signals to subsequent nerve cells. The axon is also usually highly branched at its end and forms the so-called telodendron (Kandel et al., 2000).

The coupling between two nerve cells, i.e., the place where a branch of the telodendron or axon of the predecessor neuron meets a dendrite of the successor neuron, is called a synapse. Characteristic of the structure of the synapse is a small gap with a width of about 10 to 50 nanometers, the so-called synaptic gap, which separates the presynaptic neuron, which sends signals, from the postsynaptic neuron or target cell, which receives signals.

P. Krauss, *Artificial Intelligence and Brain Research*, https://doi.org/10.1007/978-3-662-68980-6_3

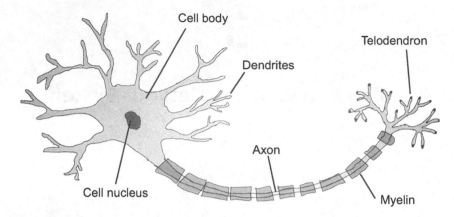

Fig. 3.1 Biological neuron. Characteristic of the neuron are two types of extensions of the cell membrane. The dendrites receive signals from other neurons and pass them on to the cell body. Action potentials are transmitted to other neurons via the axon. In many neurons, the axon is surrounded by a myelin sheath, which acts like a kind of insulation and increases the signal conduction speed

When an electrical signal, a so-called action potential, reaches the end of the presynaptic neuron, it triggers the release of chemical messengers, so-called neurotransmitters, into the synaptic gap. These neurotransmitters diffuse through the synaptic gap and then bind to receptors on the postsynaptic neuron or target cell. So, there is a chemical transmission. Depending on the number of receptors and the amount of released neurotransmitters, the synapses can be of varying strength, i.e., they can transmit a signal more or less well and thus have a varying influence on the activity of the successor neuron. In addition, there are basically two different types of synapses, which can either stimulate or inhibit the activity of the postsynaptic cell: excitatory synapses, which favor the activation of the successor neuron, and inhibitory synapses, which rather prevent the activation of the successor neuron. Formally, a real number can be assigned to each synapse, where the absolute value corresponds to the signal transmission strength and the sign corresponds to the type of synapse: positive for excitatory and negative for inhibitory. This number is called synaptic weight or short weight, often abbreviated as w. In the entirety of the synapses of the nervous system, more precisely: in the resulting network and the resulting directed information flow, all the information within a nervous system is stored, i.e., all knowledge as well as memories, learned skills, innate behaviors and reflexes, and even character traits (Kandel et al., 2000).

All signals arriving at the dendrites are transmitted at the respective synapses and weighted with the value of the synapse, which mathematically corresponds to a multiplication of the signal with the continuous synaptic weight. The information transfer from the synapse to the cell body thus uses an analog encoding. Finally, all weighted signals in the cell body of the neuron are spatially and temporally integrated into the so-called membrane potential. If this exceeds a certain threshold, the neuron becomes active and fires an action potential or spike at the axon hillock (Koch, 1999). At the same time, the membrane potential, after a short refractory period of 2 to 4 milliseconds during which no further action potential can be generated, is reset to the resting membrane potential. The action potential corresponds to a brief change in the electrical potential, triggered by ions entering the cell. Since each action potential always has about the same time course, the same strength, and the same duration of about one millisecond, the output sequence of action potentials of a neuron corresponds to a quasi-digital code. In each millisecond, either an action potential can be sent out or not. The information is represented in the temporal rate (frequency) and the exact temporal sequence of the action potentials. The emitted action potential runs along the axon until it is finally transmitted again via synapses to further subsequent nerve cells. In the nervous system, there is thus a constant change from analog encoding through chemical transmission and quasi-digital encoding through electrical transmission. It is not yet fully clarified whether the simultaneous use of both types of encoding is necessary to enable cognitive functions in the brain, or whether purely analog or purely digital encodings can also lead to comparable performances (Kolb & Whishaw, 1989, 1998; Kandel et al., 2000).

The synaptic weights of a neuron to all its successors have—although this would of course be mathematically possible and is usually the case in artificial neural networks—not different, but all the same sign. This means that a neuron only releases one type of neurotransmitter at its synapses. In contrast to artificial neural networks, there are in biological neural networks (with few exceptions) exclusively purely excitatory or purely inhibitory neurons—a property known as Dale's principle[1] (Dale, 1935; Strata & Harvey, 1999).

---

[1] No rule without exception: Actually, the effect depends on the receptor at the postsynapse. In the dopaminergic system, for example, the same transmitter can have opposite effects (Missale et al., 1998). Neurons have also been discovered that, contrary to Dale's Principle, release two different transmitters at their synapses (Vaaga et al., 2014).

Neuroplasticity

Learning is synonymous with the experience-dependent change in the connection structure of the neural networks in the brain, which is referred to as neuroplasticity. Various types of changes can be distinguished (Kolb & Whishaw, 1989, 1998; Kandel et al., 2000).

Pruning refers to the process of eliminating unused or weak synaptic connections between neurons in the brain. This thinning is an important mechanism for the development and refinement of neuronal circuits in the brain. During the development of the brain, neurons form synaptic connections on a large scale to enable a multitude of synaptic connections. However, through the process of pruning, only the most frequently used connections between neurons are reinforced and maintained, while unused or weak connections are eliminated. This leads to neural networks with more efficient and specific connections. Pruning does not only occur during the development of the brain, but can also occur in adulthood. It is believed that pruning can help to relieve the brain by removing no longer needed or redundant synaptic connections to free up resources for important connections. Moreover, pruning plays a crucial role in imprinting learning during the development of the brain (Kandel et al., 2000).

Synaptic plasticity refers to the process of strengthening or weakening the synaptic connection between neurons by enhancing or reducing activity (Kandel et al., 2000). Various mechanisms are discussed that could underlie this type of plasticity. The Hebb's rule states that the synaptic connection between two neurons is strengthened when the presynaptic neuron is active and simultaneously activates the postsynaptic neuron. In other words: "Cells that fire together wire together" (Hebb, 2005). So, if a neuron repeatedly fires and another neuron fires at the same time, then the synapse and thus the connection between the two neurons is strengthened. This is also referred to as long-term potentiation. Conversely, synaptic connections are weakened if the neurons are rarely or never active at the same time, which is also referred to as long-term depression.

The so-called Spike Timing Dependent Plasticity (STDP), on the other hand, describes how the synaptic connection between neurons changes due to the exact temporal sequence of their activity (Gerstner et al., 1996). Accordingly, the synapse is only strengthened when the presynaptic neuron is active shortly before the postsynaptic neuron, and weakened when the presynaptic neuron is active shortly after the postsynaptic neuron. In a way, STDP can be considered an extension of Hebb's rule. The dependence on the temporal sequence of neuronal activity corresponds to the consideration of causality. A neuron that already sends an action potential before the predecessor neuron did is either randomly active or was excited by another neuron. Therefore, in this case, it makes sense to weaken the synapse rather than strengthen it.

While long-term potentiation and depression act activity-dependent and thus locally on certain synapses, there is another type of synaptic plasticity, which works rather globally and is referred to as synaptic scaling (Tononi & Cirelli, 2003; De Vivo, 2017). This means that all input synapses of a neuron are weakened proportionally to their respective strength. The purpose of this process, which mainly takes place during sleep, is to prevent individual synapses from becoming increasingly stronger and the sum of all inputs of a neuron from becoming too large, causing the neuron to fire too frequently.

A similar purpose is served by intrinsic plasticity, through which the activity of a neuron can be changed (Daoudal & Debanne, 2003; Zhang & Linden, 2003). The goal is to maintain the long-term average activity level of a neuron at a medium level. From an information-theoretical point of view, a neuron that never fires is just as pointless as a neuron that constantly fires. Both processes, synaptic scaling and intrinsic plasticity, are collectively referred to as homeostatic plasticity and ensure to maintain a balance in the neural network, which is optimal for information processing (Desai, 2003; Marder & Goaillard, 2006).

Finally, neurogenesis and myelin plasticity also count as neuroplasticity. Neurogenesis refers to the brain's ability to form new neurons, which in the adult brain only occurs in the hippocampus. Myelin plasticity refers to changes in the myelin sheath, which surrounds the axons as electrical insulation and influences the speed of signal transmission.

Glia Cells

Not to be left unmentioned are the glia cells, which make up more than half of the volume of neural tissue. Although this important type of cell cannot generate action potentials itself, it still plays a significant role in the function of the nervous system. Their main functions include the myelination of neurons to accelerate signal transmission and fix the position of neurons. In addition, glia cells provide neurons with nutrients and oxygen, fight pathogens, remove dead neurons, and recycle neurotransmitters (Kandel et al., 2000).

Recent research suggests that glia cells may also play an active role in neural processing. For example, it has been found that astrocytes are involved in the regulation of neurotransmitter levels, in maintaining the correct chemical environment for neural signal transmission, and even in modulating synaptic transmission (Clarke & Barres, 2013; Sasaki et al., 2014). Microglia, on the other hand, have been associated with pruning and neural development (Schafer et al., 2012).

In some theoretical works, computer models have been simulated in which glia cells, especially astrocytes, play a role in the generation and modulation of neuronal oscillations. In these models, it is assumed that glia cells act as a kind of buffer for extracellular potassium ions (K^+) released by neurons during action potentials. By absorbing excess K^+ ions and modulating extracellular ion concentrations, glia cells could potentially regulate neuronal excitability and synchronization, which could contribute to the emergence of neuronal oscillations (Wang et al., 2012).

Conclusion

The brain has a remarkable ability to flexibly adapt to new circumstances and change its structure. The potential implications of the various types of neuroplasticity for the development of new learning algorithms in Artificial Intelligence are considerable. By studying the learning and adaptability of the brain, researchers can develop new algorithms that can learn and adapt in a similar way to the human brain. The exploration of synaptic plasticity, i.e., the strengthening or weakening of connections between neurons, has for example led to the development of artificial neural networks that mimic the structure and function of the human brain and are capable of learning from examples and adapting their behavior over time.

The exploration of structural plasticity, which involves changes in the physical structure of neurons and their connections, can also lead to the development of algorithms that can restructure in response to new data and experiences. This could lead to more flexible and adaptable AI systems that can learn from new data and adjust their behavior accordingly.

References

Clarke, L. E., & Barres, B. A. (2013). Emerging roles of astrocytes in neural circuit development. *Nature Reviews Neuroscience, 14*(5), 311–321.

Dale, H. H. (1935). Pharmacology and nerve-endings. *Proceedings of the Royal Society of Medicine, 28*, 319–332.

Daoudal, G., & Debanne, D. (2003). Long-term plasticity of intrinsic excitability: Learning rules and mechanisms. *Learning & Memory, 10*(6), 456–465.

De Vivo, L., Bellesi, M., Marshall, W., Bushong, E. A., Ellisman, M. H., Tononi, G., & Cirelli, C. (2017). Ultrastructural evidence for synaptic scaling across the wake/sleep cycle. *Science, 355*(6324), 507–510.

Desai, N. S. (2003). Homeostatic plasticity in the CNS: Synaptic and intrinsic forms. *Journal of Physiology – Paris, 97*(4–6), 391–402.

Gerstner, W., Kempter, R., Van Hemmen, J. L., & Wagner, H. (1996). A neuronal learning rule for sub-millisecond temporal coding. *Nature, 383*(6595), 76–78.

Hebb, D. O. (2005). *The organization of behavior: A neuropsychological theory.* Psychology press.

Kandel, E. R., Schwartz, J. H., Jessell, T. M., Siegelbaum, S., Hudspeth, A. J., & Mack, S. (Eds.). (2000). *Principles of Neural Science* (Vol. 4, pp. 1227–1246). McGraw-Hill.

Koch, C. (1999). *Biophysics of computation: Information processing in single neurons.* Oxford University Press.

Kolb, B., & Whishaw, I. Q. (1989). Plasticity in the neocortex: Mechanisms underlying recovery from early brain damage. *Progress in Neurobiology, 32*(4), 235–276.

Kolb, B., & Whishaw, I. Q. (1998). Brain plasticity and behavior. *Annual Review of Psychology, 49*(1), 43–64.

Marder, E., & Goaillard, J. M. (2006). Variability, compensation and homeostasis in neuron and network function. *Nature Reviews Neuroscience, 7*(7), 563–574.

Missale, C., Nash, S. R., Robinson, S. W., Jaber, M., & Caron, M. G. (1998). Dopamine receptors: From structure to function. *Physiological Reviews, 78*(1), 189–225.

Sasaki, T., Ishikawa, T., Abe, R., Nakayama, R., Asada, A., Matsuki, N., & Ikegaya, Y. (2014). Astrocyte calcium signalling orchestrates neuronal synchronization in organotypic hippocampal slices. *The Journal of Physiology, 592*(13), 2771–2783.

Schafer, D. P., Lehrman, E. K., Kautzman, A. G., Koyama, R., Mardinly, A. R., Yamasaki, R., & Stevens, B. (2012). Microglia sculpt postnatal neural circuits in an activity and complement-dependent manner. *Neuron, 74*(4), 691–705.

Strata, P., & Harvey, R. (1999). Dale's principle. *Brain Research Bulletin, 50*(5), 349–350.

Tononi, G., & Cirelli, C. (2003). Sleep and synaptic homeostasis: A hypothesis. *Brain Research Bulletin, 62*(2), 143–150.

Vaaga, C. E., Borisovska, M., & Westbrook, G. L. (2014). Dual-transmitter neurons: Functional implications of co-release and co-transmission. *Current Opinion in Neurobiology, 29*, 25–32.

Wang, F., Smith, N. A., Xu, Q., Fujita, T., Baba, A., Matsuda, T., & Nedergaard, M. (2012). Astrocytes modulate neural network activity by Ca2+-dependent uptake of extracellular K+. *Science Signaling, 5*(218), ra26–ra26.

Zhang, W., & Linden, D. J. (2003). The other side of the engram: Experience-driven changes in neuronal intrinsic excitability. *Nature Reviews Neuroscience, 4*(11), 885–900.

4

Organization of the Nervous System

Never trust something that can think for itself if you can't see where it keeps its brain!

Arthur Weasley

Modularity of the Nervous System

A fundamental organizational principle of the nervous system of more advanced species is its strong modularity. It is therefore not a gigantic homogeneous neural network, but rather consists of various functional units, which are not only anatomically distinguishable from each other, but can also operate largely independently of each other.

Another characteristic that all vertebrate nervous systems have in common is their bipolar mirror symmetry. Each anatomically distinguishable module exists twice, once in the left and once in the right half of the body. There are indeed examples in the animal kingdom of other architectures. Cephalopods, e.g. octopuses, have a nervous system with circular symmetry.

At the highest level, a distinction can be made between the central and peripheral nervous system. The central nervous system (CNS) is the main control unit of the body and consists of the brain, the brainstem, which connects the brain with the spinal cord and plays an important role in regulating vital functions such as respiration, cardiovascular and blood pressure, as well as the spinal cord. The spinal cord in turn acts as a connection between the CNS and the peripheral nervous system, which includes the autonomic

© The Author(s), under exclusive license to Springer-Verlag GmbH, DE, part of Springer Nature 2024
P. Krauss, *Artificial Intelligence and Brain Research*,
https://doi.org/10.1007/978-3-662-68980-6_4

and somatic nervous system and can be considered as an interface between the organism and the world. The autonomic nervous system controls a number of involuntary body processes such as heart rate, respiration, digestion, and pupil contraction. These processes run automatically and are not subject to conscious control. The autonomic nervous system consists of two complementary systems: Sympathetic and Parasympathetic. The sympathetic nervous system prepares the organism for activity by activating organs and body functions, thus putting the body into fight-or-flight mode. The parasympathetic, on the other hand, is active during rest and relaxation phases and contributes to the calming and recovery of the body. The autonomic nervous system responds to a variety of stimuli, including stress, physical activity, and changes in the environment. It works with other body systems to maintain homeostasis and allow the body to function optimally. While the autonomic functions cannot be consciously controlled, they can be influenced in various ways, e.g. through relaxation techniques, movement and breathing exercises. The somatic nervous system, finally, is responsible for transmitting sensory information from the sensory organs to the CNS and for forwarding motor commands from the CNS to the muscles to control voluntary movements (Fig. 4.1).

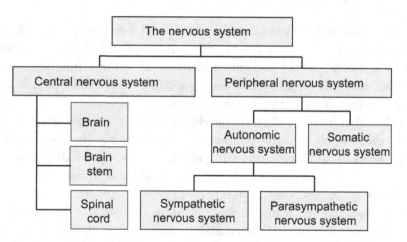

Fig. 4.1 Organization of the nervous system. The central nervous system controls the body and consists of the brain, the brainstem, and the spinal cord. The peripheral nervous system includes the autonomic and the somatic nervous system. The autonomic nervous system regulates involuntary body processes and includes the sympathetic and the parasympathetic nervous system. The somatic nervous system transmits sensory information to the CNS and forwards motor commands from the CNS to the muscles

Spinal Cord and Brainstem

The spinal cord is part of the central nervous system and runs from the brain through the vertebral canal to the lower back. It is a long, narrow strand of nerve tissue, consisting of nerve cells and nerve fibers. It plays a crucial role in transmitting motor control signals between the brain and skeletal muscles and in transmitting sensations such as pain, temperature, touch, and pressure from the sensory organs to the brain. However, the spinal cord should not be imagined as a large cable harness that simply transmits information in both directions. Rather, it is an independent control unit for the coordination of movements, which contains the circuits for controlling many reflexes as well as the programs for complex movement sequences such as walking. Reflexes are differentiated into self-reflexes and foreign reflexes (Kandel et al., 2000).

A self-reflex is an involuntary and automatic reaction of the body to a specific stimulus, which occurs without the involvement of the brain. It is a local reaction, in which sensory nerve endings in muscles or tendons transmit information to the spinal cord, which is directly connected to corresponding motor neurons, thereby triggering a motor reaction in the same muscle. Since the information from input to output is guided via only one synapse, these are also referred to as monosynaptic reflexes. A typical example is the patellar reflex, where a light blow to the tendon below the kneecap leads to a stretching of the quadriceps muscle, which in turn triggers a contraction of this muscle as a counter-reaction. Self-reflexes are important for controlling muscle tension and help to prevent muscle injuries and unwanted movements.

In contrast, a foreign reflex, also known as a polysynaptic reflex, is a reflex in which the reflex response does not originate in the organ receiving the stimulus. The reflex arc is transmitted here via several synapses from the sensory nerve endings in the body to the spinal cord, where the motor reaction is then triggered. The information is also forwarded to higher centers in the brain. However, this only serves to inform the brain retrospectively that something has just happened. The control of the foreign reflex takes place in the spinal cord, as the reaction would otherwise be much too slow. An example of a foreign reflex is the pain reflex, in which a pain stimulus on the skin (such as touching a hot stove or stepping on a sharp object) triggers an immediate withdrawal movement. Foreign reflexes serve to protect the body from possible damage and to enable a quick reaction to dangers.

The advantages of this principle of subsidiary movement control in the spinal cord are obvious. On the one hand, disturbances in the movement sequence, such as stumbling, can be reacted to very quickly, and on the other hand, it is very efficient, as the brain is relieved of recurring, trivial control tasks.

The Thalamus

The thalamus is a small but important structure deep in the center of the brain. It acts as a relay station for sensory information by processing signals from the sensory organs and forwarding them to the corresponding areas of the cerebral cortex (cortex). Only it alone is responsible for the perception and conscious processing of all sensory stimuli and the (motor) response to them. The thalamus is often referred to as the "gatekeeper" of the cerebral cortex, as it filters and regulates the flow of information to the cerebral cortex, so that only the most relevant and important sensory impressions reach consciousness. This is achieved through a network of thalamocortical loops, i.e., reciprocal connections between the thalamus and various regions of the cerebral cortex. These loops enable the activity of the cortex to be modulated and regulated by filtering out irrelevant information and amplifying important signals.

The thalamus also plays a role in regulating sleep and wakefulness as well as certain cognitive functions such as memory, attention, and language. It is connected to various other parts of the brain, including the basal ganglia, the hypothalamus, and the amygdala (parts of the so-called limbic system, which is also responsible for processing emotions), contributing to a variety of functions and behaviors.

The thalamus is not a homogeneous structure, but consists of several nuclei, each with its own function. Specific nuclei, also referred to as relay nuclei, are responsible for forwarding specific sensory information from the periphery (such as eyes, ears, skin, or taste buds) to the corresponding regions of the cerebral cortex. These nuclei act as filters, amplifying or attenuating the sensory input depending on the significance of the information. The non-specific nuclei, on the other hand, are involved in regulating the overall activity of the cortex. These nuclei receive input from various regions of the brain, including the basal ganglia, the limbic system, and the brainstem, and send projections to the cortex to modulate its activity. Non-specific thalamic nuclei play a crucial role in regulating attention, arousal, and sleep-wake cycles (Kandel et al., 2000).

The Cortex

The cerebrum, the largest part of the human brain and most higher mammals, consists of two hemispheres, which are connected by the so-called corpus callosum. This powerful bundle of fibers contains about 200 million nerve fibers, the so-called commissures. For comparison: The optic nerve, which transmits all information from the retina of the eye to the brain, consists only of 1 million nerve fibers.

The nerve cells are organized in six layers parallel to the surface of the cortex. To accommodate as large a brain surface as possible in a limited skull volume, the cerebral cortex (Cortex) is strongly folded by numerous convolutions (gyri) and furrows (sulci). This folding effect increases the number of nerve cells that can be accommodated along the surface of the cerebrum. Since the skull is subject to anatomical constraints, it is as if one were stuffing a large towel into a small cooking pot. To solve the problem, one must crumple the towel. The outer areas of the cortex contain the gray matter, which consists of the cell bodies of the approximately 16 billion neurons of the cerebrum. The rest of the cerebrum consists of the axons of the neurons, the "connecting cables", which are referred to as white matter due to their myelin sheath. With an area of about 2500 cm^2, which corresponds to about four DIN A4 pages, the cortex is only about 3 mm thick. The cortex is therefore a flat, approximately two-dimensional structure. All higher cognitive performances such as conscious perception, language, thinking, memory, movement and feelings are located in the cortex (Kandel et al., 2000), which is why the organization of the cortex will be presented in detail in the next chapter.

The Hippocampus

The Hippocampus is located in the medial temporal lobe. It owes its name to the fact that it looks like a seahorse. Together with the amygdala, it belongs to the limbic system and forms the so-called hippocampus formation with the entorhinal cortex. It receives input from virtually all regions of the cortex and is essential for the formation of new declarative and episodic memory contents, especially in the consolidation and retrieval of episodic and spatial memories, as well as for spatial navigation.

The hippocampus is considered the highest level of the cortical hierarchy, as it integrates and processes information from various brain regions before forwarding it for storage and further processing to other regions (Hawkins & Blakeslee, 2004).

Hippocampal replay is a phenomenon that occurs in the formation, consolidation, and retrieval of memories (Wilson & McNaughton, 1994). It involves the reactivation of neuronal activity patterns that were originally generated during a specific experience or event. It is believed that this repetition strengthens and stabilizes the memory trace. There are two main types of replay: awake and sleep. Awake replay occurs during rest or inactivity phases when a person is still conscious, while sleep replay occurs during non-REM sleep (deep sleep). Both types of replay are crucial for memory consolidation and the integration of new information into existing memory networks.

It is believed that the fast-learning hippocampus stores new memory contents during the day and then transfers them as a trainer to the slow-learning cortex during sleep (replay). The hippocampus thus serves as a kind of buffer or RAM (Rapid Access Memory). This is the brain's solution to the so-called stability-plasticity dilemma (Grossberg, 1982).

This dilemma is a fundamental challenge for neuronal systems, especially when it comes to learning and memory. It illustrates the need to find a balance between retaining already learned information and adapting to new experiences. Every neuronal system must be able to store and retrieve long-term memories, and at the same time be flexible enough to take in new information and adapt to changed circumstances, without forgetting what has already been learned.

A possible solution to the stability-plasticity dilemma is therefore the existence of separate memory systems in the brain, such as the hippocampus and the cortex (McClelland, 1995). The hippocampus is involved in the rapid encoding and initial consolidation of new memories, while the cortex is responsible for the slow, gradual process of memory consolidation and storage. This separation of memory systems can help ensure that new learning does not collide with already stored information, thus creating a balance between stability and plasticity.

The hippocampus also plays a key role in pattern separation (O'Reilly & McClelland, 1994) and pattern completion (McClelland, 1995). Pattern separation is the process by which similar experiences or events are represented as different memories, enabling precise retrieval and minimizing interference between memories. Pattern completion, on the other hand, is the process by which incomplete or disturbed or noisy signals can trigger the retrieval of a complete memory.

Recent studies also suggest that the hippocampus, in addition to spatial navigation (Morris et al., 1982), also enables navigation in abstract, mental spaces and is thus involved in the organization of thoughts across domains by constructing cognitive maps of any content and flexibly transferring them to new situations and contexts (Bellmund et al., 2018). If accurately replicated, this functionality could significantly improve the capabilities of our machine learning systems (Bermudez-Contreras et al., 2020).

The Basal Ganglia

The basal ganglia are a group of nuclei deep in the brain that are responsible for many important functions of motor control and learning. One of their main tasks is to act as a kind of "servomechanism" for the cortex, i.e., they assist in the regulation and fine-tuning of movements and actions initiated by the cerebral cortex (Kolb & Whishaw, 2009).

An important function of the basal ganglia is the learning and storing of complex input-output relationships. To this end, they receive input from all sensory cortex areas and send their output via the thalamus to the motor control areas of the cortex. These direct connections act like shortcuts in information transmission and enable more efficient and effective, automated movement control, as the sensory input can be converted faster and more precisely into the corresponding motor output, without having to go through the comparatively slow and complex hierarchy of the cortex areas.

Finally, the basal ganglia are also involved in a form of learning known as model-free reinforcement learning (Bar-Gad et al., 2003). This means that they learn from the results of actions and can adjust their behavior accordingly, without explicitly having to create a model of the environment. This type of learning is essential for adaptive behavior and allows us to continuously improve our actions and decisions over time. The basal ganglia are also involved in cognitive processes such as decision-making and motivation.

It is assumed that most of our movement control takes place unconsciously and automatically. For example, when learning to drive, this is a very complex and strenuous task. Among other things, you have to operate three pedals with only two feet and at the same time keep an eye on all road users and the road signs. Practically all areas of the cortex are strongly activated. But with increasing driving practice, something interesting happens. The cerebral cortex becomes less and less active. An experienced driver can drive almost unconsciously, completely automated. Especially on routine routes like the daily commute to work or on long motorway journeys, the cortex is in the so-called default mode, a resting state that the cortex always assumes when there is "nothing to do". Therefore, driving can often even be meditative for experienced drivers, as the cortex is essentially idling and one can indulge in daydreams or think about something completely different than driving. However, there is also the danger that it becomes too monotonous and boring and one falls asleep. This certainly would not have happened during the first driving lesson.[1]

The cortex becomes fully active again when a critical situation occurs. In such situations, the cortex takes over control again, and the driver focuses his full attention on the road because he has to react quickly to the unexpected. Surely you have experienced this or a similar situation yourself, whether driving or riding along. You are having a conversation during the drive, thanks to their basal ganglia the driver can use her cortex for something other than steering the car. When suddenly a child runs onto the road from behind a parked car, the driver immediately stops talking and focuses her full attention on the road.

The Cerebellum

The cerebellum (Cerebellum) is one of the oldest brain regions in terms of evolutionary history and makes up about 10% of the total volume of the brain. The cerebellum consists of two hemispheres and is divided into various lobes and nuclei. Of the total 86 billion neurons in the brain, the cerebellum contains the most neurons of all brain parts with about 69 billion neurons. In evolution, the volume of the cerebellum has increased parallel to the surface enlargement of the cortex and is most pronounced in humans (Barton & Venditti, 2014; Sereno et al., 2020).

The cerebellum is responsible for the coordination and fine-tuning of movements by integrating information from sensory systems and other brain regions and interacting with motor systems. It also supports the planning and execution of movements as well as adaptation to changed conditions and the correction of errors. In addition, the cerebellum also plays a role in

---

[1] The sequence of different phases of neural activity described here corresponds to the development of competence levels, a model from developmental psychology, starting with unconscious incompetence, conscious incompetence and conscious competence, to unconscious competence (Adams, 2011).

cognition and emotion, although its exact function in these areas is not yet fully understood (Kandel et al., 2000; Kolb & Whishaw, 2009).

One of the most important functions of the cerebellum is the temporal coordination of movements, which enables the smooth execution of complex movements such as walking, running, or playing a musical instrument. The cerebellum receives information from the spinal cord, the sense of balance, and practically all sensory cortex areas and uses this to coordinate and adjust movements. In addition to coordinating movements, the cerebellum is also responsible for correcting movement errors. If there is a discrepancy between the intended and the actual movement, the cerebellum recognizes this error and makes the necessary adjustments to correct the movement. The cerebellum also plays a crucial role in learning new motor skills. Through constant adjustment and refinement of movements, the cerebellum enables humans to improve their motor skills over time (Kandel et al., 2000; Kolb & Whishaw, 2009).

The cerebellum is divided into three main parts. The vestibulocerebellum receives information about the position and movement of the body from the vestibular system (balance organs in the inner ear), which it uses to regulate body posture and balance. It is also responsible for the precise coordination of almost all eye movements that originate from various oculomotor centers in the brainstem (Kandel et al., 2000; Kolb & Whishaw, 2009).

The spinocerebellum receives sensory signals from the spinal cord, which provide information about the position of joints and muscles, as well as continuous feedback about the movement signals sent to the spinal cord and the periphery. It also ensures fine-tuning of the movement signals and that the movement is executed as intended. This includes the complex coordination of the facial and laryngeal muscles required for speaking.

The pontocerebellum is functionally connected to the cerebral cortex. It receives input from various areas, particularly from the premotor centers in the frontal lobe (premotor cortex and supplementary motor cortex), where motor plans are formed. In the cerebellum, these are temporally precisely modulated and the planned activity of the involved muscles is coordinated. The results of these calculations are forwarded to the thalamus, from where they are finally forwarded to the motor cortex as input (Kandel et al., 2000).

There is some evidence that the cerebellum plays an important role not only for motor functions but also for cognitive processes. There are various arguments for this thesis. As already mentioned, the hemispheres of the cerebellum in humans are particularly pronounced, which in evolutionary terms goes hand in hand with the growth of the cerebral cortex and the development of human cognitive abilities. On the other hand, the cerebellum also receives a tremendous amount of information from virtually all areas of the cerebral cortex. Moreover, it has been shown that the output of the cerebellum via the thalamus reaches not only motor cortex areas but also other areas of the cortex. Interestingly, it has also been observed that certain lesions, i.e., injuries, of the cerebellum have no effect on movement coordination, suggesting that the corresponding areas must be responsible for other, non-motor functions. Finally, functional studies using modern imaging techniques have shown that the cerebellum is also activated in many cognitive tasks.

One of the most remarkable features of the cerebellum is its extremely regular strictly geometric circuit diagram (Fig. 4.2). The parallel fibers are the axons of the granule cells, which represent the most common type of neurons in the cerebellar cortex. These fibers first rise vertically, then branch into two opposite branches and finally run horizontally through the molecular layer of the cerebellum, where they form synapses with the fan-like dendrites of Purkinje cells. The parallel fibers are generally considered to be non- or very thinly myelinated, which gives them a very slow conduction speed. It is believed that this property allows the cerebellum to detect the

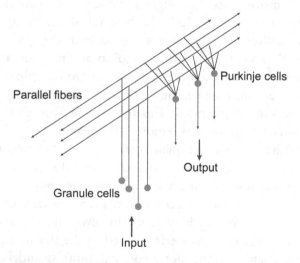

Fig. 4.2 Circuit diagram of the cerebellum. The granule cells represent the input layer of the cerebellum, their axons rise and then branch into parallel fibers. There they form synapses with the fan-like dendrites of the Purkinje cells, which function like a kind of coincidence detectors. Only when enough action potentials arrive at the same time via the parallel fibers, they are activated

smallest time differences between incoming signals. This ability is important for the precise synchronization and coordination of neuronal activity, which in turn is crucial for the role of the cerebellum in motor control, learning, and coordination (Kandel et al., 2000; Kolb & Whishaw, 2009).

Interestingly, there is a structure in the nervous system that has a very similar structure to the cerebellum, the Dorsal Cochlear Nucleus (DCN) (Oertel & Young, 2004; Bell et al., 2008; Singla et al., 2017). The DCN is one of the first processing stages of the auditory system. It is believed that here the information content of the auditory input from the cochlea (hearing snail) is determined and monitored by calculating the autocorrelation of the signal (Licklider, 1951; Krauss et al., 2016, 2017). Due to the anatomical similarity, it is obvious to assume that the cerebellum performs similar calculations and thus at least partially has a similar function to the DCN, namely to quantify and regulate the information content.

Conclusion

The brain is not a single large neural network, but on the contrary, it exhibits a very high degree of modularity. The study of the brain's modularity can provide valuable insights for the development of new AI systems, architectures, and algorithms. An important concept of machine learning is the idea of modular learning, where a complex problem is broken down into smaller, more manageable sub-problems. This is comparable to the way the brain is divided into different modules that perform specific functions.

Furthermore, studying the modularity of the brain also offers the opportunity to develop more efficient and specialized AI algorithms that can perform specific tasks with high accuracy. By breaking down a complex task into smaller subtasks and developing separate modules for each subtask, an AI system can be optimized for this specific task. This can lead to faster and more accurate performance than an AI system that tries to handle all tasks in a single module.

The modular organization of the brain allows for flexibility and adaptability, as different modules can be reconfigured and used for different tasks. Similarly, AI systems with modular architectures can be developed, allowing for easy integration of new modules and adaptation to new tasks and environments. For example, the hippocampus-cortex system could serve as a model for future AI systems to solve the stability-plasticity dilemma.

References

Adams, L. (2011). *Learning a new skill is easier said than done.* Gordon Training International.

Bar-Gad, I., Morris, G., & Bergman, H. (2003). Information processing, dimensionality reduction and reinforcement learning in the basal ganglia. *Progress in Neurobiology, 71*(6), 439–473.

Barton, R. A., & Venditti, C. (2014). Rapid evolution of the cerebellum in humans and other great apes. *Current Biology, 24*(20), 2440–2444.

Bell, C. C., Han, V., & Sawtell, N. B. (2008). Cerebellum-like structures and their implications for cerebellar function. *Annual Review of Neuroscience, 31,* 1–24.

Bellmund, J. L., Gärdenfors, P., Moser, E. I., & Doeller, C. F. (2018). Navigating cognition: Spatial codes for human thinking. *Science, 362*(6415), eaat6.

Bermudez-Contreras, E., Clark, B. J., & Wilber, A. (2020). The neuroscience of spatial navigation and the relationship to artificial intelligence. *Frontiers in Computational Neuroscience, 14,* 63.

Grossberg, S. (1982). How does a brain build a cognitive code? *Studies of Mind and Brain. Boston Studies inthe Philosophy of Science, 70.* Springer. https://doi.org/10.1007/978-94-009-7758-7_1.

Hawkins, J., & Blakeslee, S. (2004). *On intelligence.* Macmillan.

Kandel, E. R., Schwartz, J. H., Jessell, T. M., Siegelbaum, S., Hudspeth, A. J., & Mack, S. (Eds.). (2000). *Principles of neural science* (Vol. 4, pp. 1227–1246). McGraw-Hill.

Kolb, B., & Whishaw, I. Q. (2009). *Fundamentals of human neuropsychology.* Macmillan.

Krauss, P., Tziridis, K., Metzner, C., Schilling, A., Hoppe, U., & Schulze, H. (2016). Stochastic resonance controlled upregulation of internal noise after hearing loss as a putative cause of tinnitus-related neuronal hyperactivity. *Frontiers in Neuroscience, 10,* 597.

Krauss, P., Metzner, C., Schilling, A., Schütz, C., Tziridis, K., Fabry, B., & Schulze, H. (2017). Adaptive stochastic resonance for unknown and variable input signals. *Scientific Reports, 7*(1), 2450.

Licklider, J. C. R. (1951). A duplex theory of pitch perception. *The Journal of the Acoustical Society of America, 23*(1), 147–147.

McClelland, J. L., McNaughton, B. L., & O'Reilly, R. C. (1995). Why there are complementary learning systems in the hippocampus and neocortex: Insights from the successes and failures of connectionist models of learning and memory. *Psychological Review, 102*(3), 419.

Morris, R. G., Garrud, P., Rawlins, J. A., & O'Keefe, J. (1982). Place navigation impaired in rats with hippocampal lesions. *Nature, 297*(5868), 681–683.

Oertel, D., & Young, E. D. (2004). What's a cerebellar circuit doing in the auditory system? *Trends in Neurosciences, 27*(2), 104–110.

O'Reilly, R. C., & McClelland, J. L. (1994). Hippocampal conjunctive encoding, storage, and recall: Avoiding a trade-off. *Hippocampus, 4*(6), 661–682.

Sereno, M. I., Diedrichsen, J., Tachrount, M., Testa-Silva, G., d'Arceuil, H., & De Zeeuw, C. (2020). The human cerebellum has almost 80% of the surface area of the neocortex. *Proceedings of the National Academy of Sciences, 117*(32), 19538–19543.

Singla, S., Dempsey, C., Warren, R., Enikolopov, A. G., & Sawtell, N. B. (2017). A cerebellum-like circuit in the auditory system cancels responses to self-generated sounds. *Nature Neuroscience, 20*(7), 943–950.

Wilson, M. A., & McNaughton, B. L. (1994). Reactivation of hippocampal ensemble memories during sleep. *Science, 265*(5172), 676–679.

5

Organization of the Cortex

A person consists of two parts—his brain and his body. But the body has more fun.

Woody Allen

Division of the Cortex into Hemispheres and Lobes

Like practically every part of the nervous system, the cerebral cortex, or Cortex, also exists in duplicate. The two halves of the Cortex are referred to as the left and right hemisphere. Anatomically, each hemisphere can be roughly divided into five so-called lobes, which are separated from each other by deeper fissures. The division of the Cortex into lobes is anatomically and functionally significant, as each lobe can be assigned a specific primary processing area as well as a whole group of functions. On the surface of the brain are the frontal or frontal lobe, the parietal or parietal lobe, the occipital or occipital lobe, and the temporal or temporal lobe. The insular lobe or insular cortex is not visible from the outside, as it is covered by the frontal, parietal and temporal lobes (Fig. 5.1).

This anatomical division also correlates with the major functional systems of the Cortex. In the frontal lobe, which is particularly pronounced in humans, are the motor areas responsible for movement control and planning. In addition, fundamental personality traits seem to be located here at the foremost pole, the so-called orbitofrontal cortex, which lies above the

P. Krauss, *Artificial Intelligence and Brain Research*, https://doi.org/10.1007/978-3-662-68980-6_5

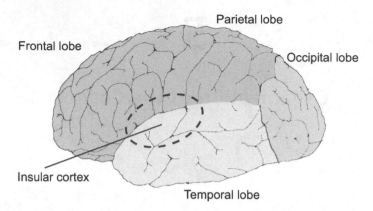

Fig. 5.1 Division of the Cortex into Lobes. Each hemisphere of the Cortex consists of five lobes: frontal lobe, parietal lobe, temporal lobe, and occipital lobe. Not visible from the outside, as it is covered by the frontal, parietal, and temporal lobes, is the fifth, the insular lobe (dotted line)

eye sockets. The parietal lobe, which contains the primary somatosensory center and areas of spatial perception, follows to the rear. The areas responsible for visual perception are located in the occipital lobe. The temporal lobe contains the auditory center as well as higher visual areas, e.g., for face recognition, and areas of long-term memory, so-called multimodal association areas. The insular cortex is the least explored and houses, among other things, the primary taste area and other areas that represent vegetative body states. Here, the primary center for basic viscerosensitivity, i.e., for information from the viscera, is also suspected.

Columns: The Radial Organization of the Cortex

In addition to the vertical organization in layers, there is a radial organization into so-called pillars or columns.

Microcolumns (Mountcastle, 1997) are smaller structures consisting of groups of about 80 neurons with similar response characteristics, which are strongly interconnected through vertical connections. Each microcolumn is selectively responsible for a specific feature of a sensory stimulus or a cognitive process. In the visual cortex, for example, a microcolumn may contain neurons that all respond to a specific orientation (angle) or a specific color of a visual stimulus at a specific location in the visual field. It is believed that microcolumns could represent the basic units of information processing in the cortex. According to this view, information is encoded in the activity of the neurons within the microcolumns and integrated across the microcolumns to form the respective output.

Due to the number of neurons in the cortex (approx. 16 billion) and the number of neurons per microcolumn (approx. 80), there is an approximate number of about 200 million microcolumns in the cortex. Interestingly, this corresponds to both the number of commissures (nerve fibers) in the corpus callosum, which connect both hemispheres, and the number of nerve fibers that connect the cortex to the cerebellum via the pons. This could suggest that each microcolumn sends its information on average exactly via one nerve fiber both into the opposite hemisphere and into the cerebellum, supporting the hypothesis that the microcolumns are the fundamental processing units of the cortex.

Macrocolumns, on the other hand, are groups of about a hundred microcolumns, which work together to process information from a specific sensory modality or cognitive function (Mountcastle, 1997). A macrocolumn consists of the representations of all possible manifestations of a certain feature at a certain location; in visual processing, for example, all possible angles or colors of a stimulus at a certain location in the visual field.

Of course, this does not mean that we can only distinguish between about a hundred different colors or angles at a given location in the visual field. Through so-called population coding, the actual number of representable feature manifestations is virtually unlimited. Population coding is a neural coding scheme that represents sensory or motor information through the activity of an entire population of neurons or microcolumns, not just the activity of individual neurons or columns. In other words, the information is represented by the combined activity of a group of neurons or microcolumns.

This coding scheme is used in the brain to represent a variety of information such as visual stimuli, auditory stimuli, and motor commands. Each neuron in the population has a preferred feature manifestation or stimulus, and its firing rate is set to respond best to this manifestation of the feature. When a stimulus is presented to the population, each neuron contributes to the overall response in proportion to its sensitivity to the stimulus.

Population coding has advantages over other coding schemes such as rate coding or temporal coding, as it can represent complex stimuli with greater accuracy and tolerates noise or fluctuations in the response of individual neurons. It also allows for the integration of information from different sensory modalities or sources, which is crucial for many cognitive functions (Averbeck et al., 2006).

Layers: The Vertical Organization of the Cortex

The cortex is remarkably uniformly structured, which is why it is also referred to as isocortex (uniform cortex). If you cut it crosswise to the surface and look at the sections under different stains in the microscope, you can recognize a vertical layer structure. The different layers are packed at different densities and consist of different types of neurons. In addition, there are various vertical and horizontal connections between the cells of the individual layers. Overall, six different layers can be distinguished, which are arranged and interconnected in a highly organized manner. Although this structure is in principle the same at any point in the cortex, there are also differences. Depending on where you examine, the relative thickness and cell density of the individual layers vary in the cross-section.

Brodmann Areas

In the early twentieth century, the psychiatrist and neuroanatomist Korbinian Brodmann systematically examined many sections from all regions of the cortex and found that the cortex can be divided into different areas, each characterized by a unique combination of cell types, densities, and layers (Brodmann, 1910). Brodmann published a system for mapping the human brain based on its cytoarchitecture, i.e., the organization of cells in the various brain regions. The Brodmann areas named after him are numbered according to their location in the brain, from the primary sensory and motor areas in the upper central part of the brain to the more complex association areas in the front part of the brain. The numbering system ranges from 1 to 52, with some numbers omitted because it was later found that the areas were duplicates or combinations of other areas (Fig. 5.2).

Interestingly, these structural anatomical differences correlate with functional differences. This means that each Brodmann area, which externally differs from its neighbors under the microscope, actually performs a different function. For example, it has been found that visual information is processed in Brodmann areas 17, 18, and 19 in the back of the brain, while auditory information is processed in areas 41 and 42 at the top edge of the temporal lobe. Area 4 corresponds to the primary motor cortex, which sends control commands to the skeletal muscles, and areas 1, 2, and 3 correspond to the primary sensory cortex, which is responsible for processing tactile stimuli from the body surface.

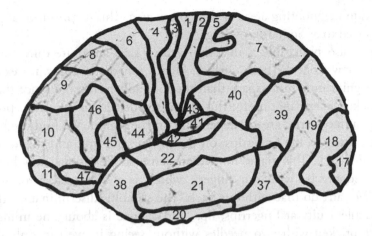

Fig. 5.2 Brodmann Areas. Each Brodmann area is defined by a unique combination of cell types, cell densities, and layers and is associated with different functions and behaviors. For example, Brodmann area 4 in the primary motor cortex is responsible for controlling voluntary movements, while Brodmann area 17 in the primary visual cortex is responsible for processing visual information

However, the actual number of cortex areas is significantly larger than the division into 52 Brodmann areas suggests (Amunts & Zilles, 2015; Coalson et al., 2018; Gordon et al., 2016). Using multimodal MRI data from the Human Connectome Project (Elam et al., 2021) and an objective, semi-automatic neuroanatomical approach, 180 areas per hemisphere have recently been identified that differ in their architecture, function, connectivity, or topography (Glasser et al., 2016).

But even this is likely to be only a preliminary, approximate number. The practical difficulty of dividing the cortex into meaningful maps or areas, and the fundamental difficulty of assigning specific functions to these maps that are neither too narrowly defined nor too general, is described by David Poeppel as the *Maps Problem* (division into maps) and *Mapping Problem* (assignment of functions) (Poeppel, 2012).

Maps in the Head

Another special feature is the map-like organization of these areas. If we look at the primary somatosensory area, i.e., the cortex area where information from the touch and touch receptors of the skin surface is processed, we notice that there is a system. Input from neighboring areas of the skin is also processed in neighboring areas of this cortex area. This organizational principle is referred to as Somatotopy (Fig. 5.3).

What is also noticeable is that the relative sizes are quite different from what we know from the body surface. Hand, tongue, and lips, for example, take up significantly more space than the foot and even more space than the entire back. The explanation for this is that the density of touch receptors on the body surface varies greatly depending on the region. We find the highest receptor density in the fingertips, on the tongue and lips, and the lowest on the legs and back. This makes immediate sense, as we can feel very high-resolution and precise with our fingers, which is also the basis for reading Braille. We can't do that with our back. The stimulus discrimination threshold or tactile acuity at fingertips, lips, and tongue is about one millimeter: If we are pricked with two needles without seeing it, we can only distinguish whether we were pricked with one or two needles from a distance of less than one millimeter. In contrast, tactile acuity on legs and back is even

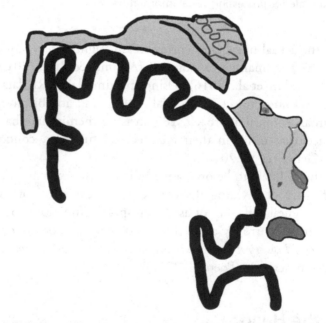

Fig. 5.3 Somatosensory Homunculus. Projection of the body surface onto the primary sensorimotor cortex area. The relative size of the body parts is greatly distorted and is based on the respective density of touch and tactile receptors. The representations of hand, tongue, and lips take up significantly more space than those for foot and back

several centimeters. High tactile acuity means that a lot of information from the corresponding body region reaches the cortex, which in turn requires a lot of processing capacity, i.e., a relatively large area of the corresponding area. In contrast, only a relatively small area is needed for processing for the back, where the receptor density is lowest.

The primary motor area, which sends control commands to the muscles, behaves analogously. We can move our fingers very finely tuned and precisely, but not so much our foot. Accordingly, the areas for controlling hands and face are relatively large compared to the areas that control the muscles of the feet and toes. Try playing the piano with your feet!

The primary visual area (visual cortex) also follows this topographic organizational principle. Here we speak of retinotopy, as the input comes from the retina. The retina is structured in such a way that the spatial relationships between the different parts of the visual field are preserved. This means that adjacent parts of the retina correspond to adjacent parts of the visual field. When the information from the optic nerve reaches the visual cortex in the brain, it is organized in a similar way. Adjacent regions of the visual cortex correspond to adjacent regions of the visual field, creating a "map" of the visual world. Again, this map is not a perfect one-to-one representation of the visual field. Rather, there are distortions and irregularities in the way different parts of the visual field are represented in the brain. The fovea, the point of sharpest vision in the relatively small central area of the retina, is greatly overrepresented in the primary visual cortex. Although the fovea only makes up about 1 percent of the total area of the retina, about half of the primary visual cortex is responsible for processing information from the fovea.

The reason for this strong overrepresentation lies again in the high sensitivity and visual acuity of the fovea. Since the fovea contains a high density of photoreceptors and is responsible for our sharpest vision, it provides the brain with the most detailed and accurate information about the visual world. Therefore, the brain assigns a larger proportion of its processing resources to the fovea to ensure that this information is processed with the greatest possible accuracy. In contrast, the peripheral regions of the visual field are represented in the visual cortex in a highly compressed form.

The last example refers to the auditory system, which is responsible for the perception of sounds and speech. Here too, there is a map-like structure, the tonotopy, where adjacent frequencies are processed at adjacent locations of the primary auditory area. The acoustic input is broken down into its individual frequency components at the basilar membrane of the cochlea (hearing snail), with similar frequencies represented at adjacent locations on the basilar membrane. This tonotopy continues in all further processing stations of the auditory pathway up to the primary auditory cortex area.

As is well known, most bats use ultrasonic echo location to locate their prey and obstacles in their environment even in darkness. They can then derive information about the position and movement of objects in their environment based on the runtime and frequency shift of the echo. The frequency range of the ultrasonic waves emitted by the bats is therefore particularly important for them and is represented in their auditory system with high resolution, i.e., they can recognize the smallest frequency differences in this range and it therefore takes up a disproportionately large space in their primary auditory cortex area, which is why, by analogy with the human visual system, we also speak of an "auditory fovea".

In fact, the map-like organization can be found in all sensory modalities and in the motor system, where it is most obvious in the primary areas, as these areas are closest to the outside world, so to speak. The primary sensory areas represent the input areas of the cortex, i.e., they receive their inputs from the sensory organs, while the primary motor area is the output area of the cortex, which sends its output to the muscles and thus quasi to the outside world. In contrast, all other areas receive their input from other cortex areas and also send their output back to other cortex areas. Since the structure of the cortex is fundamentally the same everywhere, it can be assumed that the fundamental functional principles are also the same in all cortex areas. This means that each area creates a map of its input, with similar input represented at adjacent locations and more frequent and therefore more important input taking up more space on the map. However, as the respective area gets further away from the outside world, the representations become increasingly abstract and the map-like organization becomes harder to understand.

The Canonical Circuit of the Cortex

The six layers of the cortex are interconnected in a specific way. Although fundamentally all possible combinations of connections exist, some are much more pronounced than others. From this fact, the following pattern for the flow of information within a cortical area emerges: The input from the sensory organs (in primary sensory areas) or from other, lower cortical areas first reaches layer 4. From there, the information is forwarded to layers

2/3 and finally to layer 5/6, which serves as the output layer of the cortex, from where the information is transmitted back to the thalamus or to lower cortical areas as feedback. In addition, fibers from layers 2/3 project into further cortical areas. Depending on which layer is the source and target, three different types of connections can be distinguished between two cortical areas: ascending (bottom-up, feed-forward), descending (top-down, feedback), and horizontal (lateral) connections (Kolb & Whishaw, 2009; Kandel et al., 2000; Imam & Finlay, 2020) (Fig. 5.4).

Hierarchy of Cortical Areas

By analyzing the type of connections between the cortical areas, a kind of circuit diagram of the information flow through the areas can be reconstructed. It turns out that the areas of the cortex are hierarchically-parallel organized, i.e., at each hierarchy level there is usually not just one, but several areas, with the number of areas per level also increasing with increasing hierarchy level. Interestingly, it turns out that the hippocampus is at the top of the hierarchy of sensory areas (Kolb & Whishaw, 2009; Felleman & Van Essen, 1991; Van Essen et al., 1992) (Fig. 5.5).

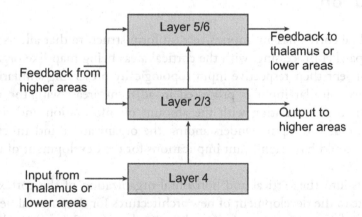

Fig. 5.4 Information flow in the cortical layers. The six layers of the cortex are specifically interconnected. Input from sensory organs or hierarchically lower cortical areas reaches layer 4, is forwarded to layers 2/3, and finally reaches the output layers 5/6. From there, feedback is sent back to the thalamus or hierarchically lower cortical areas. In addition, fibers from layers 2/3 project into hierarchically higher cortical areas or across the corpus callosum into homologous areas on the other hemisphere. Feedback from higher areas ends in layers 2/3 and 5/6

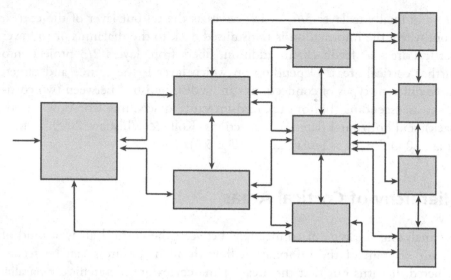

Fig. 5.5 Hierarchical-parallel organization of the cortex. The analysis of the connections between cortical areas allows the reconstruction of a circuit diagram of the information flow. The areas are hierarchically and parallel arranged, with several areas present at each level and the number of areas per level increasing with higher hierarchy level

Conclusion

The cerebral cortex has an impressively uniform structure that allows a hierarchical-parallel processing, with the cortical areas being map-like organized. They represent their respective input topologically preserving similarity and importance. Similar input is processed in adjacent areas, with the relative size of the area correlating with the amount of information and thus the importance of the input. Understanding the organization and function of the cortex could have significant implications for the development of new AI architectures.

Insights into the vertical and horizontal organization of the cortex could contribute to the development of new architectures for deep neural networks to make information processing in different layers and between different areas of the network more efficient. In addition, the study of the functioning of micro- and macro-columns as well as population coding in the cortex could contribute to developing new approaches for information processing and representation in artificial neural networks that may be more robust and efficient. The analysis of canonical circuits and the flow of information in the cortex could contribute to the development of new algorithms and learning strategies for artificial neural networks that consider both local and global information and respond better to feedback.

Finally, understanding the role of the hippocampus in the hierarchy of sensory areas and its involvement in information processing and integration could contribute to developing new models of artificial intelligence that are better able to process spatio-temporal information and build a long-term memory.

References

Amunts, K., & Zilles, K. (2015). Architectonic mapping of the human brain beyond Brodmann. *Neuron, 88*(6), 1086–1107.

Averbeck, B. B., Latham, P. E., & Pouget, A. (2006). Neural correlations, population coding and computation.*Nature Reviews Neuroscience, 7*(5), 358–366.

Brodmann, K. (1910). *Feinere Anatomie des Grosshirns* (pp. 206–307). Springer.

Coalson, T. S., Van Essen, D. C., & Glasser, M. F. (2018). The impact of traditional neuroimaging methods on the spatial localization of cortical areas. *Proceedings of the National Academy of Sciences, 115*(27), E6356–E6365.

Elam, J. S., Glasser, M. F., Harms, M. P., Sotiropoulos, S. N., Andersson, J. L., Burgess, G. C., ... & Van Essen, D. C. (2021). The human connectome project: A retrospective. *NeuroImage, 244*, 118543.

Felleman, D. J., & Van Essen, D. C. (1991). Distributed hierarchical processing in the primate cerebral cortex. *Cerebral Cortex, 1*(1), 1–47.

Glasser, M. F., Coalson, T. S., Robinson, E. C., Hacker, C. D., Harwell, J., Yacoub, E., ... & Van Essen, D. C. (2016). A multi-modal parcellation of human cerebral cortex. *Nature, 536*(7615), 171–178.

Gordon, E. M., Laumann, T. O., Adeyemo, B., Huckins, J. F., Kelley, W. M., & Petersen, S. E. (2016). Generation and evaluation of a cortical area parcellation from resting-state correlations. *Cerebral Cortex, 26*(1), 288–303.

Imam, N., & Finlay, B. L. (2020). Self-organization of cortical areas in the development and evolution of neocortexs. *Proceedings of the National Academy of Sciences, 117*(46), 29212–29220.

Kandel, E. R., Schwartz, J. H., Jessell, T. M., Siegelbaum, S., Hudspeth, A. J., & Mack, S. (Eds.). (2000). *Principles of neural science* (Vol. 4, pp. 1227–1246). McGraw-Hill.

Kolb, B., & Whishaw, I. Q. (2009). *Fundamentals of human neuropsychology.* Macmillan.

Mountcastle, V. B. (1997). The columnar organization of the neocortex. *Brain: A Journal of Neurology, 120*(4), 701–722.

Poeppel, D. (2012). The maps problem and the mapping problem: Two challenges for a cognitive neuroscience of speech and language. *Cognitive Neuropsychology,* *29*(1–2), 34–55.

Van Essen, D. C., Anderson, C. H., & Felleman, D. J. (1992). Information processing in the primate visual system: An integrated systems perspective. *Science,* *255*(5043), 419–423.

6

Methods of Brain Research

One only sees what one already knows and understands.

Johann Wolfgang von Goethe

Imaging Techniques: Watching the Brain Think

Imaging techniques are methods used to create images of the brain or other body structures. Imaging techniques play a crucial role in brain research as they allow researchers to collect and analyze detailed information about the brain. With these techniques, brain activity can be measured during various tasks, and it can be traced how different brain regions interact with each other. Imaging techniques can also help detect and investigate brain diseases and disorders such as stroke, dementia, and epilepsy. Overall, imaging techniques are an essential part of brain research and contribute to expanding our understanding of the brain and how it works.

CT

The Computed Tomography (CT) uses X-rays to create detailed cross-sectional images of the brain. X-ray images of the head are taken from various angles and reconstructed into a three-dimensional image of the brain on a computer. The resulting image shows the structures of the brain, including

P. Krauss, *Artificial Intelligence and Brain Research*,
https://doi.org/10.1007/978-3-662-68980-6_6

the ventricles, the skull, and the blood vessels. CT imaging is a common method for diagnosing neurological diseases such as stroke, traumatic brain injury, and brain tumors. However, CT imaging does not provide information about the functional activity of the brain, which can be obtained with other imaging techniques such as PET, fMRI, EEG, and MEG (Ward, 2015; De Groot & Hagoort, 2017).

PET

The Positron Emission Tomography (PET) visualizes the metabolic activity of cells and tissues in the body. A small amount of a radioactive substance, a so-called radiotracer, is injected into the body. The radiotracer emits positrons, positively charged particles that interact with electrons in the body. When a positron encounters an electron, they annihilate each other and produce gamma rays, which are detected by the PET scanner. The PET scanner records these gamma rays and creates a three-dimensional image of the brain's metabolic activity from them. However, since the brain is constantly active, meaningful data with PET can only be generated by subtracting two images. Typically, an image is taken during a specific cognitive task or stimulus, and another image of the brain's background activity is taken, and then the difference image is calculated (Ward, 2015; De Groot & Hagoort, 2017).

MRI

The Magnetic Resonance Imaging (MRI), also known as nuclear magnetic resonance imaging, uses the property of hydrogen nuclei to align along magnetic field lines to generate detailed images. Short magnetic pulses stimulate the aligned nuclei to emit electromagnetic waves, which are detected and used to determine the distribution of hydrogen, mainly in the form of water. Functional Magnetic Resonance Imaging (fMRI) uses the fact that the red blood pigment affects the magnetic signal of the hydrogen nuclei differently, depending on whether it has bound oxygen or not. Comparing measurements under different stimulation conditions allows changes in blood oxygen saturation (Blood Oxygenation Level Difference, BOLD) to be derived, which serve as an indirect measure of the change in blood flow to a specific brain area. The assumption behind this is that active brain areas require more oxygen and are therefore more heavily perfused. The BOLD signal builds up slowly and reaches a maximum about six to ten seconds after the stimulus begins, before it slowly decreases again. Compared to MEG and EEG, nuclear magnetic resonance imaging has a lower temporal resolution of about one recording per second (1 Hz), but a much higher spatial resolution in the range of about one cubic millimeter (Menon & Kim, 1999), i.e., the activity of the brain can be measured in about one million so-called voxels[1] simultaneously (Ward, 2015; De Groot & Hagoort, 2017).

EEG and MEG

The Electroencephalography (EEG) is a non-invasive method for measuring the electrical activity of the brain using electrodes placed on the scalp. The various patterns of the brain's electrical activity are used to diagnose neurological diseases such as epilepsy, sleep disorders, brain tumors, and head injuries, and are also important in research for studying brain functions and behavior. The EEG is known for its extremely high temporal resolution, as it can perform up to 100,000 measurements per second (100 kHz). However, the spatial resolution is limited, and the electrical fields are strongly attenuated by the brain tissue. With so-called high-density EEG systems, brain activity can be measured simultaneously in 128 channels distributed over the scalp. The EEG is best suited for measuring the activity of the cerebral cortex, especially the gyri, as they are in close proximity to the skull bone. For measuring the activity of deeper brain regions, the MEG is better suited (Ward, 2015; De Groot & Hagoort, 2017).

The Magnetoencephalography (MEG) is also a non-invasive method and is used to measure the magnetic activity of the brain using superconducting magnetometers (Hämäläinen et al., 1993), so-called SQUIDs (Super Conducting Quantum Interference Devices). Similar to the EEG, the MEG can be used to diagnose neurological diseases such as epilepsy or brain tumors, but also in research for studying brain functions and behavior. The MEG also has an extremely high temporal resolution, as brain activity can be recorded at up to 100,000 measurements per second (100 kHz). The spatial resolution is slightly better than that of the EEG (approx. 250 magnetometers distributed over the scalp). Since magnetic fields are practically not attenuated or distorted by brain tissue, the MEG is particularly well suited for measuring activity in the sulci of the cortex and also in deeper

---

[1] A voxel, from volume pixel, is the three-dimensional analogue to a pixel (picture element).

brain regions below the cerebral cortex (subcortical regions). MEG and EEG therefore complement each other well and can be used simultaneously in combined M/EEG measurements (Ward, 2015; De Groot & Hagoort, 2017).

Event-related Potentials and Fields

Event-related potentials (ERPs) (engl. Event-Related Potentials, ERP) are a measure of brain activity that can be measured with EEG while the test person performs a certain task or is confronted with certain stimuli (Handy, 2005; Sur & Sinha, 2009). ERPs are temporally aligned with the presentation of the stimulus or event and are only visible when averaged over many (at least 50) measurement repetitions. This is because the noise on the actual ERPs is ten to a hundred times greater than the actual signal. ERPs represent the neuronal activity associated with the cognitive or sensory processing of an event. ERPs are usually characterized by their polarity, latency, and amplitude. The polarity refers to whether the electrical potential recorded at the scalp is positive or negative relative to a reference electrode. The latency is the time interval between the stimulus presentation and the occurrence of the peak of the ERP waveform. The amplitude reflects the strength or size of the electrical potential recorded at the scalp. ERPs are often used in cognitive neuroscience to investigate cognitive processes such as attention, memory, language, perception, and decision-making. They can provide insights into the neuronal mechanisms underlying these processes and can also be used as biomarkers for various neurological and psychiatric disorders. Event-related fields (Event-Related Fields, ERF) correspond to the magnetic analogue of the ERPs measured with MEG (Ward, 2015; De Groot & Hagoort, 2017).

Intracranial EEG

Intracranial Electroencephalography (iEEG) is an invasive technique for measuring brain activity (Parvizi & Kastner, 2018). Electrodes are placed directly on or in the brain, usually during surgery, to record electrical activity. This allows for very high spatial resolution, as the electrodes are placed in specific regions of the brain, often in close proximity to the area of interest. iEEG also has a very high temporal resolution, allowing electrical signals to be recorded at a rate of up to several thousand times per second (Ward,

2015; De Groot & Hagoort, 2017). Patients being treated for drug-resistant epilepsy have iEEG electrodes implanted for diagnostic purposes a few weeks before surgery to resect their epilepsy focus, to determine which brain regions need to be spared during surgery to preserve important functions such as language. These patients often participate in neuropsychological studies, as the data obtained in this way are extremely rare and valuable.

Conclusion

Today, various techniques are available to measure the structure and activity of the brain. Each method has specific advantages and disadvantages, and no method is perfect. Ideally, complementary methods such as MEG and EEG or fMRI and EEG are used simultaneously to get a more complete picture of brain activity. But even then, we are still far from being able to read the brain in the spatiotemporal resolution that would be necessary to capture the exact temporal activity course of each neuron and each synapse. This is one reason why computer models of brain function are absolutely necessary. In contrast to the brain, simulated models offer the crucial advantage that all internal parameters and variables can be read out at any time with any desired accuracy.

References

De Groot, A. M., & Hagoort, P. (Eds.). (2017). *Research methods in psycholinguistics and the neurobiology of language: A practical guide*. Wiley.

Hämäläinen, M., Hari, R., Ilmoniemi, R. J., Knuutila, J., & Lounasmaa, O. V. (1993). Magnetoencephalography – Theory, instrumentation, and applications to noninvasive studies of the working human brain. *Reviews of Modern Physics, 65*(2), 413.

Handy, T. C. (Eds.). (2005). *Event-related potentials: A methods handbook*. MIT press.

Menon, R. S., & Kim, S. G. (1999). Spatial and temporal limits in cognitive neuroimaging with fMRI. *Trends in Cognitive Sciences, 3*(6), 207–216.

Parvizi, J., & Kastner, S. (2018). Promises and limitations of human intracranial electroencephalography. *Nature Neuroscience, 21*(4), 474–483.

Sur, S., & Sinha, V. K. (2009). Event-related potential: An overview. *Industrial Psychiatry Journal, 18*(1), 70.

Ward, J. (2015). *The student's guide to cognitive neuroscience*. Psychology Press.

7

Memory

Long-term memory is the result of permanent changes brought about by the growth of new synaptic connections.

Eric Kandel

Memory as an Information Processing System

Memory is a crucial aspect of human cognition, as it allows the brain to encode, store, and retrieve information over an extended period (Milner et al., 1998; Kandel, 2007). The recollection of past events forms the basis for the development of language, social relationships, and personal identity. Without memory, human experience would be severely limited.

Memory is organized as a multi-stage information processing system. In the first stage, the sensory organs absorb information from the environment. The subsequent modality-specific neural processing acts as ultra-short-term or sensory memory. The preprocessed information from the sensory systems is then transferred to short-term and possibly working memory. In this phase, the incoming sensory information is temporarily stored and possibly integrated and processed with information from other modalities and cognitive systems. From there, the information may be consolidated and transferred to long-term memory (Atkinson & Shiffrin, 1968). Long-term memory is responsible for storing information over longer periods up to the entire lifespan. In this phase, memories become part of the personal biography and contribute to the development of personality and self-image.

© The Author(s), under exclusive license to Springer-Verlag GmbH, DE, part of Springer Nature 2024
P. Krauss, *Artificial Intelligence and Brain Research*,
https://doi.org/10.1007/978-3-662-68980-6_7

The ability to store and retrieve information over an extended period is also crucial for controlling future actions. The information stored in memory forms the basis for decisions and allows the individual to draw on past experiences to influence current actions. Thus, memory plays a decisive role in controlling an organism's behavior and responses to its environment (Squire, 1987, 1993; Mandler, 1967).

Sensory Memory

The sensory memory is the first stage in the memory's information processing system. It stores information received through the sensory organs only for a very short period, usually less than a second. Sensory memory is crucial for us to perceive and process the world around us. Depending on the sensory modality, different types of sensory memory are distinguished, including iconic, echoic, and haptic memory.

Iconic memory refers to visual stimuli and allows an image to be briefly retained in the mind's eye (Sperling, 1963). This type of memory plays a crucial role in tasks such as reading, where the eyes need to move quickly to process the text. Echoic memory, on the other hand, refers to auditory stimuli and allows sounds or words to be retained for a short time, which forms the basis for the ability to process language and communication. Finally, haptic memory refers to tactile stimuli and allows sensations of physical touch to be stored for a short time. This type of memory is important for people to navigate their environment and interact with objects in their surroundings.

Sensory memory is believed to correspond to short-term neural activity in specific sensory brain regions such as the visual or auditory cortex. The exact neural correlates of sensory memory vary depending on the type of sensory information being stored, and research continues into how this information is processed and maintained in the brain (Gazzaniga et al., 2006).

The partial-report paradigm, introduced by George Sperling in 1963 (Sperling, 1963), is a method for investigating sensory memory. In the classic version, subjects are presented with a matrix of letters or numbers for a very short time, usually about 50 milliseconds. Immediately afterwards, a tone or cue is given indicating which row of the grid the participants should name. The subjects are then asked to repeat the letters or numbers from the indicated row. Sperling found that participants were able to name almost all the letters in the indicated row if the tone or cue was given immediately after the grid disappeared, even though the participants did not know which row was relevant before the matrix was presented. This suggests that sensory memory must have a relatively large storage capacity, as apparently all rows of the matrix can be stored for a short period of time. However, if the cue to the row was delayed by a few hundred milliseconds, participants usually could only remember very few letters from the corresponding row. This, in turn, suggests that sensory memory fades quickly.

Sperling also introduced another variant of the paradigm, the full-report paradigm, in which the test subjects are asked to remember all the letters or numbers of the grid, not just those in a particular row. In general, however, participants are then only able to remember a few letters or numbers from the entire matrix, regardless of whether the cue is given immediately or later. This suggests that conscious access to sensory memory has a limited capacity, which is smaller than the actual storage capacity of sensory memory.

The partial-report paradigm is widely used in research on sensory memory and attention and has provided important insights into the nature of these processes. It has also been used to investigate visual and auditory perception as well as the encoding and retrieval of memories (Fig. 7.1).

Short-term Memory

Short-term memory refers to the ability to temporarily store a small amount of information for a few seconds to a minute without altering it. It is based on short-term neural activity in multimodal brain regions, particularly in the prefrontal, parietal, and temporal cortex (Fuster & Alexander, 1971; Funahashi et al., 1989). The involvement of the thalamus, basal ganglia, and cerebellum is also discussed.

Short-term memory is essential for daily life as it allows us, for example, to remember important details such as phone numbers, addresses, or directions. However, short-term memory has a very limited capacity. This is often associated with the "magic number" seven. This goes back to a publication by George Miller from the 1950s (Miller, 1956). There he suggested that the average number of elements or "chunks" that can be stored in short-term memory is seven plus or minus two. Accordingly, most people should be able to store five to nine units of information at a given time. However, recent research suggests that the actual number may be somewhat lower, with an average capacity of only about four chunks (Cowan, 2001; Luck & Vogel, 1997; Rouder et al., 2008). This means that if we try to remember more than these four to five objects or contents, we do it very poorly and we will probably forget some of them.

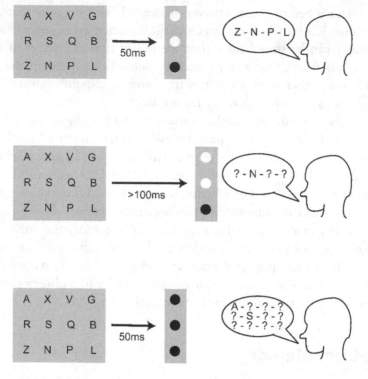

Fig. 7.1 Partial-Report Paradigm. Top: Sensory memory has a relatively large storage capacity, as almost all letters in the indicated row can be named when the cue is given immediately. Middle: However, it fades quickly, as shown when the cue is given with a delay. Bottom: The full-report paradigm shows that conscious access to sensory memory has a limited capacity, which is smaller than the storage capacity of sensory memory itself

Fortunately, there are methods to increase the capacity of short-term memory. An effective strategy is so-called chunking, where individual pieces of information are grouped into larger blocks. For example, if we try to remember a phone number, we can divide it into groups of three or four digits to make it easier to remember.

Give it a try. Please remember this number:

3471892341

Did you manage? If you're like most people, including the author, you'll find it difficult or even impossible to remember the number this way.

Now try to memorize this number:

347 189 2341

It should have been much easier for you now. The difference is that in the first case the number consists of ten digits that you have to remember, while in the second case only three chunks, each consisting of three to four digits, need to be remembered.

As we have just seen, this chunking interestingly works even when the individual chunks have no special meaning. However, if the chunks have a meaningful significance, the storage capacity of short-term memory can be significantly further increased, as it can then be applied recursively, which is also referred to as hierarchical chunking.

Try to remember these two characters:

你 好

Unless you happen to speak and especially read Mandarin, it is virtually impossible to quickly memorize these two characters, which by the way mean "You good" or "Hello". This is because in reality you would have to remember not just two, but actually—depending on the count—form, size, position, and orientation of about twenty individual objects, the strokes.

If you had learned the characters before because you learned Mandarin, then instead of the twenty strokes, you would now only have to remember two individual objects, the characters for "you" and "good" in short-term memory, or if you know that both together mean as much as "Hello", even only a single object.

Working Memory

Working memory is a theoretical concept that plays a central role in cognitive psychology, neuropsychology, and cognitive neuroscience. It is an important cognitive system that allows us to temporarily store and manipulate information. This system is essential for many higher cognitive functions such as logical reasoning, problem-solving, decision-making, and behavioral control.

Working memory consists of several subcomponents (Baddeley & Hitch, 1974), each of which has its own function (Fig. 7.2). The central executive component is responsible for controlling attention and coordinating information processing. Another component is the phonological loop, which is responsible for the temporary storage of verbal information. It can be further divided into the articulatory loop, which is responsible for the subvocal repetition of verbal information, and the acoustic buffer, which is responsible

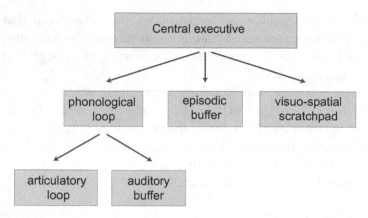

Fig. 7.2 Multicomponent model of working memory. Working memory consists of several subcomponents. The central executive controls attention and coordinates information processing. The phonological loop stores verbal information, while the episodic buffer integrates information from different modalities and the visuospatial sketchpad enables the temporary storage and manipulation of visuospatial information

for the temporary storage of linguistic information. The episodic buffer is another component of working memory, serving as a temporary storage system for the integration of information from different modalities, such as visual and auditory information. Finally, the visuospatial sketchpad is responsible for the temporary storage and manipulation of visuospatial information. We use this system, for example, when mentally rotating an object.

The neural correlates of short-term and working memory are the result of numerous complex brain activities involving a multitude of brain regions (Goldman-Rakic, 1995). Studies have shown that the central executive component of working memory is associated with activities in the prefrontal cortex and the basal ganglia (Hikosaka et al., 1989; Graybiel, 1995). The phonological loop is associated with activities in the left hemisphere, particularly in the parietal and temporal lobes. The visuospatial sketchpad is associated with activity in the posterior parietal lobe and the prefrontal cortex region (Milner et al., 1985). The episodic buffer, on the other hand, is associated with activity in the medial temporal region and the hippocampus. The thalamus is also believed to play an important role in maintaining working memory through the control and modulation of feedback loops with the cortex (Watanabe & Funahashi, 2012). Finally, the cerebellum is also attributed a (yet not fully understood) function in the control and regulation of short-term memory (Ivry, 1997; Desmond & Fiez, 1998).

Long-term Memory

Long-term memory is essentially based on synaptic plasticity (Bailey & Kandel, 1993) and allows us to store and retrieve information over longer periods—from hours to decades. Two categories of this type of memory are distinguished: declarative (explicit) memory and non-declarative, procedural (implicit) memory (Eichenbaum & Cohen, 2004).

Declarative memory refers to the conscious retrieval of information and can be further subdivided into semantic memory, which contains general factual knowledge, and episodic memory, which contains specific events and their sequence (Tulving, 1972, 1983). These do not necessarily have to be events or episodes experienced by oneself, but can also be movies watched or the plot of books read. A special case of episodic memory is autobiographical memory, which contains memories of specific events from one's own life (Conway & Pleydell-Pearce, 2000). The medial temporal lobe and the hippocampus are important brain regions involved in the formation and retrieval of declarative long-term memories (Milner, 1970; Zola-Morgan & Squire, 1986); Squire, 1987, 1992; Gazzaniga et al., 2006). As we have already seen, it is particularly the case that the hippocampus acts as a kind of intermediate storage for declarative memories (Scoville & Milner, 1957; Penfield & Milner, 1958) and transfers its information to the cortex during deep sleep. This process is also referred to as memory consolidation (Squire, 1992; Conway & Pleydell-Pearce, 2000).

Non-declarative, procedural memory, on the other hand, refers to our unconscious memory for skills, procedures, and motor abilities such as playing the piano, swimming, or cycling. This type of memory often manifests itself in our behavior and does not need to be consciously retrieved. The cerebellum and the basal ganglia are key regions of the brain involved in the formation and retrieval of procedural long-term memory (Graybiel, 1995; Desmond & Fiez, 1998).

Conclusion

In summary, memory is an integral part of human cognition that allows us to encode, manipulate, store, and retrieve information over time. Understanding how memory functions as an information processing system is crucial to understanding its role in our experience and behavior.

The study of human memory has significant implications for the development of Artificial Intelligence. If we understand how human memory works, we can develop AI systems that learn and process information more efficiently than previous systems. One of the most important areas of research in this context is the development of working memory models for AI systems. Working memory is crucial for many cognitive tasks, including problem-solving and decision-making. By integrating a working memory into AI systems, their abilities to handle complex tasks and make more accurate predictions could be significantly improved.

In addition, the study of long-term memory has implications for the development of Artificial Intelligence that is supposed to learn and adapt over time. Just as humans store and retrieve information over long periods, these systems could be designed to store and retrieve large amounts of data and use this information to improve their performance over time. This is particularly important in areas such as natural language processing or real image processing, where the systems constantly need to adapt to new data and contexts.

Since human memory is a complex, multi-layered system that is not yet fully understood, the development of AI systems that simulate human memory still poses a major challenge.

References

Atkinson, R. C., & Shiffrin, R. M. (1968). Human memory: A proposed system and its control processes. In *Psychology of learning and motivation* (Vol. 2, pp. 89–195). Academic: Elsevier.

Baddeley, A. D., & Hitch, G. (1974). Working memory. In *Psychology of learning and motivation* (Vol. 8, pp. 47–89). Academic, Elsevier.

Bailey, C. H., & Kandel, E. R. (1993). Structural changes accompanying memory storage. *Annual Review of Physiology, 55*(1), 397–426.

Conway, M. A., & Pleydell-Pearce, C. W. (2000). The construction of autobiographical memories in the self-memory system. *Psychological Review, 107*(2), 261.

Cowan, N. (2001). The magical number 4 in short-term memory: A reconsideration of mental storage capacity. *Behavioral and Brain Sciences, 24*(1), 87–114.

Desmond, J. E., & Fiez, J. A. (1998). Neuroimaging studies of the cerebellum: Language, learning and memory. *Trends in Cognitive Sciences, 2*(9), 355–362.

Eichenbaum, H., & Cohen, N. J. (2004). *From conditioning to conscious recollection: Memory systems of the brain* (No. 35). Oxford University Press on Demand.

Funahashi, S., Bruce, C. J., & Goldman-Rakic, P. S. (1989). Mnemonic coding of visual space in the monkey's dorsolateral prefrontal cortex. *Journal of Neurophysiology, 61*(2), 331–349.

Fuster, J. M., & Alexander, G. E. (1971). Neuron activity related to short-term memory. *Science, 173*(3997), 652–654.

Gazzaniga, M. S., Ivry, R. B., & Mangun, G. R. (2006). *Cognitive Neuroscience. The biology of the mind.* Norton.

Goldman-Rakic, P. S. (1995). Cellular basis of working memory. *Neuron, 14*(3), 477–485.

Graybiel, A. M. (1995). Building action repertoires: Memory and learning functions of the basal ganglia. *Current Opinion in Neurobiology, 5*(6), 733–741.

Hikosaka, O., Sakamoto, M., & Usui, S. (1989). Functional properties of monkey caudate neurons. III. Activities related to expectation of target and reward. *Journal of Neurophysiology, 61*(4), 814–832.

Ivry, R. (1997). Cerebellar timing systems. *International Review of Neurobiology, 41*, 555–573. PMID: 9378608.

Kandel, E. R. (2007). *In search of memory: The emergence of a new science of mind.* Norton.

Luck, S. J., & Vogel, E. K. (1997). The capacity of visual working memory for features and conjunctions. *Nature, 390*(6657), 279–281.

Mandler, G. (1967). Organization and memory. In *Psychology of learning and motivation* (Vol. 1, pp. 327–372). Academic, Elsevier.

Milner, B., Petrides, M., & Smith, M. L. (1985). Frontal lobes and the temporal organization of memory. *Human Neurobiology, 4*(3), 137–142.

Miller, G. A. (1956). The magical number seven, plus or minus two: Some limits on our capacity for processing information. *Psychological Review, 63*(2), 81.

Milner, B. (1970). Memory and the medial temporal regions of the brain. *Biology of Memory, 23*, 31–59.

Milner, B., Squire, L. R., & Kandel, E. R. (1998). Cognitive neuroscience and the study of memory. *Neuron, 20*(3), 445–468.

Penfield, W., & Milner, B. (1958). Memory deficit produced by bilateral lesions in the hippocampal zone. *AMA Archives of Neurology & Psychiatry, 79*(5), 475–497.

Rouder, J. N., Morey, R. D., Cowan, N., Zwilling, C. E., Morey, C. C., & Pratte, M. S. (2008). An assessment of fixed-capacity models of visual working memory. *Proceedings of the National Academy of Sciences, 105*(16), 5975–5979.

Scoville, W. B., & Milner, B. (1957). Loss of recent memory after bilateral hippocampal lesions. *Journal of Neurology, Neurosurgery, and Psychiatry, 20*(1), 11.

Sperling, G. (1963). A model for visual memory tasks. *Human Factors, 5*(1), 19–31.

Squire, L. R. (1987). *Memory and brain.* Oxford University Press.

Squire, L. R. (1992). Memory and the hippocampus: A synthesis from findings with rats, monkeys, and humans. *Psychological Review, 99*(2), 195.

Squire, L. R., Knowlton, B., & Musen, G. (1993). The structure and organization of memory. *Annual Review of Psychology, 44*(1), 453–495.

Tulving, E. (1972). Episodic and semantic memory. In *Organization of memory* (pp. 381–402). Academic, Elsevier.

Tulving, E. (1983). *Elements of episodic memory*. Oxford University Press.

Watanabe, Y., & Funahashi, S. (2012). Thalamic mediodorsal nucleus and working memory. *Neuroscience & Biobehavioral Reviews, 36*(1), 134–142.

Zola-Morgan, S., & Squire, L. R. (1986). Memory impairment in monkeys following lesions limited to the hippocampus. *Behavioral Neuroscience, 100*(2), 155.

8

Language

Language is, by its nature, a communal thing, that is, it never expresses the exact,
but a compromise—that which is common to you, me, and everyone else.

Thomas Ernest Hulme

How does Man Come to Language?

The question of how man comes to language is one of the oldest scientific puzzles. According to the Greek historian Herodotus, the Egyptian Pharaoh Psammetichus tried to fathom the origin of language 2500 years ago. To this end, he conducted an experiment with two children, whom he handed over to a shepherd as newborns to feed and care for them, but who was instructed not to speak to them. Psammetichus hypothesized that the first word of the infants would be spoken in the original language of all humans. When one of the children cried "bekos", which was the sound of the Phrygian word for "bread", Psammetichus concluded that Phrygian was the original language of all humans (Rawlinson & Wilkinson, 1861). Behind this cruel experiment was obviously the assumption that humans are born with innate words and their meanings and that this original language is somehow deformed or overwritten during individual development and first language acquisition.

Nowadays, of course, it is clear that words and meanings are not innate, but are learned during language acquisition, and that there is no causal relationship between the sound pattern and the meaning of a word

© The Author(s), under exclusive license to Springer-Verlag GmbH, DE, part of Springer
Nature 2024
P. Krauss, *Artificial Intelligence and Brain Research*,
https://doi.org/10.1007/978-3-662-68980-6_8

(de Saussure, 1916). However, it is still very controversial to what extent language skills are innate or need to be learned (Goodluck, 1991).

Noam Chomsky's Universal Grammar

According to the theory of universal grammar, every human being has an innate, genetically determined language ability, which, for example, distinguishes between different word classes such as nouns and verbs, which should not only facilitate, but even make it possible for children to learn to speak (Chomsky, 2012, 2014).

The theory states that there are certain universal principles and rules that underlie all human languages, and that these principles are hard-wired in our brains from birth. The concept of universal grammar was first proposed by the linguist Noam Chomsky in the 1950s. Chomsky held the view that children are born with an innate knowledge of the basic grammatical structure of language, which enables them to quickly learn the specific rules of their mother tongue when confronted with it.

One of the basic assumptions of universal grammar is that all languages have a set of basic structural features, such as the use of verbs, nouns, and adjectives, and the ability to form sentences with subject, verb, and object. It is believed that these features reflect the innate cognitive abilities of the human brain and are present in all languages regardless of their specific linguistic features.

Cognitive Linguistics: Usage-Based Approaches

In contrast to Universal Grammar and other approaches in linguistics that focus on formal rules and structures, Cognitive Linguistics investigates how language is processed and represented in our brain. It also explores how language and cognition interact and how they are influenced by factors such as culture and social interaction. In so-called usage-based approaches *(usage-based approaches)* to language acquisition, a profound relationship between language structure and language use is assumed (Goldberg, 1995, 2003, 2019; Tomasello, 2005; Langacker, 2008). In particular, it is assumed that contextual mental processing and mental representations have the cognitive capacity to capture the complexity of actual language use at all levels (Bybee et al., 1994; Hopper & Bybee, 2001; Bybee, 2013; Diessel et al., 2019; Schmid, 2020).

In the usage-oriented approaches, the importance of a child's experience with language for the language acquisition process is emphasized. These theories assume that children learn language by observing and imitating the language usage patterns in their linguistic environment. In particular, the idea that there is a fixed, innate language knowledge that children possess from birth, as postulated by the theory of Universal Grammar, is rejected. Instead, it is argued that children build their linguistic knowledge through the accumulation of language usage patterns.

An important concept in usage-based approaches is the idea of construction learning. According to this theory, children learn to use language by acquiring certain language usage patterns or constructions that they hear in the language around them. Constructions are form-meaning pairs and can range from individual words to simple phrases and sentences to more complex structures and are learned through repeated contact and use. Cognitive Linguistics views language as a complex system of constructions that are formed on the basis of experience and perception and shape our linguistic abilities and our understanding of language. This implies that, contrary to the assumptions of classical linguistics, there is no strict separation between words (lexicon) and grammar rules (syntax), but rather a lexicon-syntax continuum, i.e., there can also be all possible transitional forms from words on the one hand and rules on the other. An example of this would be so-called collocations.

A collocation is a combination of words that often occur together and are perceived as a natural unit in language use. These combinations are often idiomatic and cannot always be explained by the meaning of their individual words. The use of a collocation can contribute to making a sentence sound more fluent and natural. A collocation is particularly present when there are several possibilities to express a fact through different word combinations, but speakers or writers tend to prefer a certain combination, otherwise the statement loses its naturalness. For example, "strong soup" and "strong coffee" sound much more natural than "strong soup" and "strong coffee", although grammatically and semantically all combinations would be possible. With a strict separation of words and grammar rules, such phenomena can only be poorly explained, as then all alternative word combinations should occur approximately equally frequently.

Usage-based approaches also emphasize the role of social interaction and communication in language acquisition. They argue that language is learned through social interaction with other speakers and that children's understanding of language is shaped by the communicative functions that language fulfills in social situations.

Language in the Brain

Language is probably the most complex cognitive ability of humans, involving all cognitive subsystems: attention, movement planning and control, memory, perception, etc. Accordingly, language processing in the brain consists of complex, interconnected neural networks that are responsible for various aspects of language, including language comprehension, language production, and language acquisition.

The primary areas of the brain responsible for language processing are located in the left hemisphere, particularly in the regions of the frontal, temporal, and parietal lobes (Pulvermüller, 2002). These areas work together to process various aspects of language, e.g., phonology (the sounds of language), syntax (the rules for combining words into sentences), semantics (the meaning of words and sentences), and pragmatics, i.e., the use of language in social contexts (Herbst, 2010).

The Broca area is a brain region in the left frontal lobe and is considered the motor speech center. It is named after the French neurologist Paul Broca, who was the first to recognize the importance of this area for language processing. The Broca area is primarily involved in language production, particularly in the ability to produce language and form grammatically correct sentences. Damage to this area can lead to a condition known as Broca's aphasia, characterized by difficulties in forming coherent sentences, although the affected person still has a relatively good understanding of language. Recent studies also suggest that the Broca area may play a role in other cognitive functions beyond language processing, such as working memory, attention, and decision-making (Kemmerer, 2014).

The second important area in language processing is the Wernicke area, which is also considered the sensory language center and is located in the posterior part of the left temporal lobe near the auditory areas (Wernicke, 1874). It is named after the German neurologist Carl Wernicke, who was the first to recognize that the Wernicke area is primarily involved in language comprehension, particularly in the ability to understand and interpret spoken and written language. Damage to this area can lead to a condition known as Wernicke's aphasia, characterized by difficulties in language comprehension, but also in the production of coherent language. However, recent studies also suggest that the Wernicke area may play a role in other cognitive functions beyond language processing, such as visual perception and memory (Kemmerer, 2014). These recent findings on the function of the Broca and Wernicke areas point to two problems that neuroscientist

David Poeppel refers to as the *Maps Problem* and the *Mapping Problem* (see also Chap. 5).

One of the most important processes in language comprehension is the so-called parsing, in which the brain breaks down a sentence into its components and assigns a meaning to each part. This requires working memory, attention, and the ability to quickly process and integrate information from various sources.

Language production involves a similar series of processes, in which the brain first generates a message and then encodes it into a sequence of words and grammatical structures. This process also requires the integration of various information sources, including semantic knowledge, syntactic rules, and social context (Kemmerer, 2014).

Language acquisition is a complex process in which various cognitive and neural mechanisms such as attention, memory, and social perception interact. It is assumed that children are particularly sensitive to language during critical developmental phases and that contact with language during these phases is crucial for the acquisition of language competence.

According to a new model proposed for understanding the functional anatomy of language, the early stages of language processing take place in the auditory areas located in the temporal lobe of the cortex on both sides of the brain. Analogous to the visual system, which has a ventral "where" and a dorsal "what" path of processing, language processing in the cortex also divides into a ventral and a dorsal path (Hickok & Poeppel, 2004, 2007, 2016). While the "what" path is responsible for assigning sounds to meanings, the "where" path maps sounds onto motor-articulatory representations (where is the sound produced).

Of course, not only the cortex is involved in the processing of language. The thalamus, the basal ganglia, and the cerebellum also play an important role in language perception and production.

The thalamus, as a sensory relay station in the brain that receives and processes incoming sensory information before it is forwarded to the corresponding cortical areas for further processing, is also involved in the preprocessing of auditory information, including speech sounds.

The basal ganglia are involved in the planning and execution of speech movements. They also play a role in the formation of grammatical structures and in the selection of appropriate words during language production.

And finally, the cerebellum, traditionally associated with motor coordination and balance, is also involved in cognitive processes such as language processing. It is responsible for the coordination of speech movements as well as for the timing of language production and plays an important role in learning (foreign) languages.

Conclusion

The question of how humans acquire language has occupied science for millennia. The study of human language and language development has deepened our understanding of how language is acquired, processed, and represented in the brain. Theories such as Chomsky's Universal Grammar and usage-based approaches offer different perspectives on language acquisition, while neuroscience research expands our knowledge of the underlying brain mechanisms.

In the context of Artificial Intelligence and large language models like GPT-4, these insights provide important knowledge for the development and improvement of language processing systems. The study of human language abilities and language development can help optimize the architecture and learning mechanisms of such models and improve their ability to understand and generate natural language.

Future research in linguistics and neuroscience could help further close the gap between human and artificial language processing. The integration of insights from various disciplines such as cognitive linguistics, neuroscience, and AI research could lead to even more powerful and human-like language models. Furthermore, such models could deepen our understanding of human language and cognition by serving as tools for investigating linguistic phenomena and cognitive processes.

References

Bybee, J. L., Perkins, R. D., & Pagliuca, W. (1994). *The evolution of grammar: Tense, aspect, and modality in the languages of the world* (Vol. 196). University of Chicago Press.

Bybee, J. L. (2013). Usage-based theory and exemplar representations of constructions. In Th. Hoffmann & G. Trousdale (Eds.), *The Oxford Handbook of Construction Grammar* (online edn, 16 Dec. 2013), Oxford Academic. https://doi.org/10.1093/oxfordhb/9780195396683.013.0004. Accessed 9 Aug 2023.

Chomsky, N. (2012). On the nature, use and acquisition of language. In *Language and meaning in cognitive science* (pp. 13–32). Taylor and Francis.

Chomsky, N. (2014). *Aspects of the theory of syntax* (Vol. 11). MIT press.

De Saussure, F. (1916). Nature of the linguistic sign. *Course in General Linguistics, 1,* 65–70.

Diessel, H., Dabrowska, E., & Divjak, D. (2019). Usage-based construction grammar. *Cognitive Linguistics, 2,* 50–80.

Goldberg, A. E. (1995). *Constructions: A construction grammar approach to argument structure.* University of Chicago Press.

Goldberg, A. E. (2003). Constructions: A new theoretical approach to language. *Trends in Cognitive Sciences, 7*(5), 219–224.

Goldberg, A. E. (2019). *Explain me this: Creativity, competition, and the partial productivity of constructions.* Princeton University Press.

Goodluck, H. (1991). *Language acquisition: A linguistic introduction.* Blackwell.

Herbst, T. (2010). *English Linguistics.* De Gruyter Mouton.

Hickok, G., & Poeppel, D. (2004). Dorsal and ventral streams: A framework for understanding aspects of the functional anatomy of language. *Cognition, 92*(1–2), 67–99.

Hickok, G., & Poeppel, D. (2007). The cortical organization of speech processing. *Nature Reviews Neuroscience, 8*(5), 393–402.

Hickok, G., & Poeppel, D. (2016). Neural basis of speech perception. In G. Hickok & S. L. Small (Eds.), *Neurobiology of Language* (pp. 299–310). Academic.

Hopper, P. J., & Bybee, J. L. (2001). Frequency and the emergence of linguistic structure. In *Typological Studies in Language* (pp. 1–502). John Benjamins Publishing Company. http://digital.casalini.it/9789027298034.

Kemmerer, D. (2014). *Cognitive neuroscience of language.* Psychology Press.

Langacker, R. (2008). Cognitive grammar as a Basis for Language Instruction. In *BookHandbook of Cognitive Linguistics and Second Language Acquisition* (Vol. 1). Imprint Routledge.

Pulvermüller, F. (2002). *The neuroscience of language: On brain circuits of words and serial order.* Cambridge University Press.

Rawlinson, H. C., & Wilkinson, J. G. (1861). *The history of Herodotus* (Vol. 1). D. Appleton & Co. New York.

Schmid, H. J. (2020). *The dynamics of the linguistic system: Usage, conventionalization, and entrenchment.* Oxford University Press.

Tomasello, M. (2005). *Constructing a language: A usage-based theory of language acquisition.* Harvard University Press.

Wernicke, C. (1874). *Der aphasische Symptomencomplex: Eine psychologische Studie auf anatomischer Basis.* Cohn & Weigert.

9

Consciousness

Consciousness is not a journey upward, but a journey inward.

Bernard Lowe

An age-old mystery

For more than 2000 years, the question of understanding consciousness has been at the center of interest for many philosophers and scientists. Modern philosophy distinguishes between an easy and a hard problem (Chalmers, 1995). While the easy problem consists of explaining the function, dynamics, and structure of consciousness, the hard problem is to explain whether and why any physical system, be it a human, an animal, a fetus, a cell organoid, or an AI (Bayne et al., 2020), is conscious and not unconscious at all. Throughout history, many different perspectives have been proposed, ranging from the pessimistic view of "Ignorabimus"—which means as much as "We will never know"[1]—to more optimistic mechanistic ideas that even aim at the construction of an artificial consciousness. These different views have led to ongoing debates and discussions about the nature of consciousness and whether it can ultimately be understood or not.

---

[1] Emil du Bois-Reymond made this statement at the 45th annual meeting of German natural scientists and doctors in 1872 in his lecture on the limits of scientific knowledge, referring to the relationship between brain processes and subjective experience.

P. Krauss, *Artificial Intelligence and Brain Research*, https://doi.org/10.1007/978-3-662-68980-6_9

At the core, it is about the mind-body problem or, more modernly expressed, the problem of the relationship between the brain and the mind. The core question is: How do mental, mental states and processes relate to physical states and processes?

On the one hand, we have the brain, which is a material, physical object that can be measured and studied. On the other hand, we have the mind, which consists of subjective, conscious experiences that cannot be observed or measured in the same way as physical properties. This raises two fundamental questions. The first is whether the mind is physical or something completely different. Some philosophers believe that the mind is a non-physical entity that exists independently of the body, while others believe that the mind is merely a byproduct of the physical processes in the brain. The second question concerns the causal relationship between mind and brain: How do physical processes in the brain lead to subjective conscious experiences? Some philosophers believe that there is a one-to-one correspondence between the physical processes in the brain and the mental processes in the mind. Others believe that mental processes cannot be reduced to physical processes and that the mind is more than just the activity of the brain.

Closely related to this is the concept of qualia. These are subjective experiences from the first-person perspective that we have when we perceive or interact with the world. These experiences include sensations such as color, taste, and sound, but also more complex experiences like feelings and thoughts. Qualia are often considered ineffable, which means that they cannot be fully captured or conveyed by language or other forms of representation. They cannot, therefore, be studied from the "outside", i.e., from the third-person perspective. This has led some philosophers to claim that qualia represent a special kind of phenomena that cannot be reduced to or explained by physical or objective properties of the world.

Monism and Dualism

Over the course of history of philosophy, an innumerable variety of ideas and concepts have been developed to solve the mind-body problem, all of which can essentially be divided into one of two basic views: monism and dualism.

Monism is the idea that there is only one kind of state or substance in the universe. This concept goes back to the ancient Greek philosopher Aristotle, who believed that mind and body relate to each other as form and matter.

For Aristotle, the various forms in the physical world were simply different physical states, and there was no non-physical or mental substance beyond the physical world.

Dualism, on the other hand, assumes that both mental and physical substances are possible. This thought was first developed by Plato, who believed that mind and body exist in two separate worlds. According to Plato's idea, the mind was part of the ideal world of forms, it was immaterial, non-extended, and eternal. The idea of an ideal circle, for example, exists in the mind as a perfect concept, although no physical circle in the material world can ever truly correspond to it. In contrast, the body belongs to the material world, it is extended and perishable. Concrete physical circles could be found in the material world, but they would always be imperfect and change.

What is it Like to be a Bat?

The main problem in analyzing consciousness is its subjectivity. Our mind is capable of perceiving and processing our own states of consciousness. Through induction, we are also able to attribute conscious processes to other people. However, as soon as we try to imagine being another species, as philosopher Thomas Nagel describes in his groundbreaking essay "What is it like to be a bat?" (Nagel, 1974), we immediately fail to consciously pursue this experience.

The basic idea is to imagine what it would be like to experience the world as a bat, from the bat's own subjective perspective.

"...imagine that one has webbing on one's arms,which enables one to fly around at dusk and dawn catching insects in one'smouth; that one has very poor vision, and perceives the surrounding world by asystem of reflected high-frequency sound signals; and that one spends the dayhanging upside down by one's feet in an attic. In so far as I can imagine this(which is not very far), it tells me only what it would be like for me tobehave as a bat behaves. But that is not the question. I want to know what itis like for a bat to be a bat." (Nagel, 1974).

Nagel argues that it is impossible for us to fully understand the subjective experience of a bat or any other creature, as we are fundamentally limited by our own human perspective. We can study the behavior and physiology of bats, but we will never truly know what it is like to perceive the world through echolocation or to experience the unique sensory and cognitive processes that a bat uses to navigate and interact with its environment in total darkness.

According to Nagel's view, consciousness is not just about objective facts or physical processes, but also about subjective experiences, i.e., what it is like to be a particular organism. He suggests that we need to develop a new kind of science that takes into account subjective experience if we hope to fully understand the nature of consciousness.

He further argues that the subjective nature of consciousness undermines any attempt to decipher it with objective and reductionist means, i.e., the means of natural sciences. He believes that the subjective character of experience cannot be explained by a system of functional or intentional states.

He concludes that consciousness cannot be fully understood if its subjectivity is ignored, as it cannot be explained reductionistically, as it is a mental phenomenon that cannot be reduced to materialism. Nagel concludes with the assertion that physicalism, while not wrong, is also only incompletely understood, as it lacks the characterization of subjective experience. This, in turn, is a necessary prerequisite for understanding the mind-body problem.

At its core, he thus represents a dualistic perspective, much like Plato two and a half millennia before him.

Consciousness as a Useful Illusion?

Philosopher Daniel Dennett, on the other hand, holds a monistic view and contradicts Nagel's claim that the consciousness of the bat is fundamentally inaccessible to us. He justifies his view by stating that all important features of the bat's consciousness can be observed from the outside. For example, it is clear that bats cannot recognize objects that are more than a few meters away, as their biological echolocation system has a limited range. Dennett believes that, analogously, all aspects of a bat's subjective experience can be discovered through further scientific investigations (Dennet, 1991).

Furthermore, Dennett holds the view that consciousness is not a unified phenomenon, but rather a collection of mental processes that constantly change and interact with each other. He describes consciousness as a kind of useful illusion that arises from the functioning of the brain (Cohen & Dennet, 2011). Similarly, neuroscientist Anil Seth equates our subjective experience with a constantly newly generated hallucination, which is continuously compared with the input from the sensory organs from the world (Seth, 2021). For Dennett, the subjective experience of consciousness is an emergent property of brain activity and not a separate thing that exists in itself. He describes consciousness as a "virtual machine" that the brain creates to help us navigate the world. Dennett also holds the view that consciousness does not take place in a single central location in the brain, but rather arises from a series of parallel, distributed processes that constantly generate and update multiple different drafts of our experience (Dennet, 1991).

Dennett's perspective on consciousness emphasizes the importance of understanding the underlying mechanisms of the brain in order to comprehend the nature of subjective experience. He is skeptical of traditional dualistic views of consciousness that postulate a separate immaterial soul or mind, and instead emphasizes the importance of studying consciousness as a product of the physical brain.

Limits of the Philosophy of Mind

A disadvantage of exploring consciousness with philosophical means is that we will never be able to investigate the underlying physical, biological, or information-theoretical processes. By thinking alone, we will not be able to open the black box.

While philosophy plays an important role in expanding our knowledge and understanding of the nature of consciousness, its ability to ask more comprehensive questions, to look at the bigger picture, and to explore the relationships between different areas, rightly makes it an essential part of cognitive science. However, philosophy can only help us gain new insights and a more comprehensive understanding of consciousness through collaboration with empirical sciences.

Neural Correlates of Consciousness

In neuroscience, it is currently assumed that consciousness functions as a kind of user interface (Seth & Bayne, 2022). It comes into play whenever we have to find our way in new situations, anticipate possible future events, plan actions, consider different scenarios, and choose between them (Graves et al., 2011). While the philosophical tradition from Aristotle to Descartes considered consciousness to be an exclusively human phenomenon, modern neuroscience tends to view consciousness as a gradual phenomenon that can fundamentally also occur in animals (Boly et al., 2013).

The neural correlates of consciousness (NCC) refer to specific patterns of neural activity that are believed to be associated with conscious experience

(Crick & Koch, 2003; Koch, 2004). In other words, NCCs are the brain processes necessary for the occurrence of consciousness. The concept of NCC is based on the idea that there is a close relationship between neural activity and conscious experience. When we perceive something consciously, there are certain patterns of neural activity that are always associated with this experience. By investigating these patterns of neural activity, we hope to gain a better understanding of the mechanisms underlying conscious experience. There are various approaches to identifying NCCs. One approach is to compare brain activity during conscious and unconscious states such as sleep or anesthesia. Another approach is to investigate the changes in neural activity that occur when a person perceives a stimulus, e.g., a visual or auditory cue.

Francis Crick[2] and Christof Koch proposed that brain waves with a frequency range between 30 and 100 cycles per second, so-called gamma oscillations, play a crucial role in the emergence of consciousness (Crick & Koch, 1990). Koch further developed this concept and investigated the neural correlates of consciousness in humans (Tononi & Koch, 2008; Koch et al., 2016). Accordingly, activity in the primary visual cortex is essential for conscious perception, but not sufficient, as activity in the hierarchically higher cortical areas of the visual system is more closely correlated with the various aspects of visual perception and damage to these areas can selectively impair the ability to perceive certain features of stimuli (Rees et al., 2002)—a phenomenon referred to as agnosia. We will return to this later in this chapter.

Furthermore, Koch suggests that the precise timing or synchronization of neural activity may be much more important for conscious perception than simply the extent of neural activity. Recent studies using imaging techniques on visually triggered activity in parietal and prefrontal cortex regions seem to confirm these hypotheses (Boly et al., 2017).

Based on the idea of the neural correlates of consciousness, a kind of measurement procedure was even developed, which aims to determine the degree of consciousness (Perturbational Complexity Index) e.g., in comatose patients, independently of the activity associated with sensory and motor processing (Seth et al., 2008; Casali et al., 2013; Casarotto et al., 2016). While this approach may be suitable for quantifying the complexity of neural activity (Demertzi et al., 2019), it does not provide information about the underlying neural circuits and the "algorithms" implemented in them.

---

[2] This is, by the way, the same Francis Crick who deciphered the double helix structure of DNA in 1953 together with James Watson.

Integrated Information Theory

Building on observations of the neural correlates of consciousness, Giulio Tononi proposed the Integrated Information Theory (IIT) (Tononi et al., 2016). This theoretical framework attempts to explain how conscious experience arises from the physical activity of the brain (Tononi, 2004). According to the theory, consciousness is not simply the result of the activity of individual neurons or brain regions, but arises from the integrated activity of the brain as a whole. Tononi argues that an information-processing system can only be conscious if the information is integrated into a unified whole. In other words, it must be impossible to decompose the system into quasi-independent parts, as these parts would otherwise appear as two separate conscious entities (Koch, 2013).

The theory further assumes that the key to consciousness is the degree of information integration in the brain. This degree of integration is supposed to be quantifiable by a mathematical quantity called "phi", which represents the amount of causal information generated by interactions between different parts of the brain. The higher the phi value, the more integrated the information processing in the brain and the more conscious the system should be. The Integrated Information Theory assumes that consciousness is a fundamental property of certain physical systems that are capable of influencing themselves. Accordingly, any system with a phi value above a certain threshold would be conscious to some degree.

Theoretical physicist Max Tegmark even generalized Tononi's framework further from a consciousness based on neural networks to any physical quantum systems. He proposed that consciousness can be understood as a state of matter with pronounced information processing capabilities—thus quasi an additional state of aggregation besides solid, liquid and gaseous. He suggested the name "perceptronium" for this type of conscious matter (Tegmark, 2014).

Is the Brain a Quantum Computer?

Where we are already at physics, we want to turn to another question—at least for physicists—that is obvious.

Although there is a large consensus that consciousness can be understood as a form of information processing (Cleeremans, 2005; Seth, 2009; Reggia et al., 2016; Grossberg, 2017; Dehaene et al., 2014, 2017), there is a disagreement about what the appropriate level of description is (Kriegeskorte & Douglas, 2018). Physicist Roger Penrose and physician Stuart Hameroff hypothesized that the brain is a kind of quantum computer. In their view, quantum calculations take place in the so-called microtubules of the cell skeleton of neurons (Penrose, 1989, 1994; Hameroff & Penrose, 1996a, b; Hameroff, 2001; Hameroff et al., 2002). However, this view is highly controversial.

Max Tegmark and Christof Koch argue that the brain can be understood within a purely neurobiological framework, without having to resort to quantum mechanical properties: Quantum computations, which are based on the phenomenon of entanglement of (sub-)atomic particles, require on the one hand that the qubits (quantum mechanical analogue to the bit) are perfectly isolated from the rest of the system, while on the other hand a coupling of the system with the outside world is necessary for the input, control and output of the computations (Nielsen & Chuang, 2002). Due to the wet and warm nature of the brain, all these operations introduce disturbances in the form of noise into the computations, leading to decoherence of the quantum states and thus making quantum computations impossible. Furthermore, they argue that the molecular machines of the nervous system, such as the pre- and postsynaptic receptors, are so large that they can be treated as classical physical systems and not as quantum systems. They conclude that cognition can be fully understood within the theoretical framework of neural networks, without having to consider quantum phenomena. (Tegmark, 2000; Koch & Hepp, 2006, 2007; Koch, 2013).

Global Workspace Theory

As an alternative and competing theory to the Integrated Information Theory, Baars introduced the concept of a virtual global workspace (*Global Workspace*) to describe consciousness in the 1990s, which arises from the networking of various brain areas (Newman & Baars, 1993; Baars, 1994; Baars & Newman, 1994; Baars, 2017). This global workspace allows the integration and exchange of information between different specialized brain regions, and it is assumed that it is responsible for the emergence of conscious experiences by enabling the selective transmission of information relevant to the individual's current goals and needs.

In the Global Workspace Theory (GWT), conscious perception is compared to a spotlight that illuminates the contents of the global workspace and makes it consciously perceptible (Mashour et al., 2020). The contents of the workspace can be sensory information, thoughts, feelings, and memories. The theory states that the more comprehensively and intensively information is processed in the global workspace, the more likely it is to reach the threshold of consciousness. This means that stimuli that are particularly noticeable, novel, or emotionally significant reach consciousness more quickly.

The idea of the global workspace was later taken up and further developed by Stanislas Dehaene, and neural correlates were proposed that underlie it, especially distributed networks of prefrontal, parietal, and sensory cortex areas (Dehaene et al., 1998; Dehaene & Naccache, 2001; Dehaene & Changeux, 2004; Sergent & Dehaene, 2004; Dehaene et al., 2011, 2014) (Fig. 9.1).

Starting from the implications of this theory, namely that *"consciousness arises from specific types of information processing processes that are physically realized by the hardware of the brain"* (Dehaene et al., 2017), Dehaene argues that

"a machine equipped with these processing capacities would behave as if it had consciousness; it would, for example, know that it sees something, express its confidence in what it has seen, communicate it to others, have hallucinations when its control mechanisms fail, and even experience the same perceptual illusions as humans" (Dehaene et al., 2017) .

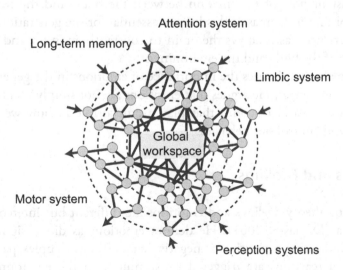

Fig. 9.1 Global workspace. Various neural processes interact with each other and exchange information. The global workspace acts as a central "stage" where consciously perceived information is processed and coordinated

The feeling of what happens!—Damasio's Model of Consciousness

The neuroscientist and psychologist Antonio Damasio has created the world's largest database of brain lesions. It contains information about the location of the lesion and the associated cognitive deficits of tens of thousands of patients, which he has collected over decades. Based on this database and his observations, he has developed his own theory of consciousness.

His view of consciousness can be summarized in the idea that consciousness arises from the interaction between our biological self and the environment with which we interact (Damasio, 1999, 2014; Damasio & Meyer, 2009). For Damasio, consciousness is not a single thing or a single process, but a complex and dynamic phenomenon that arises from the interaction between different brain regions and other body systems. In his view, consciousness is closely associated with emotions, feelings, and subjective experiences.

Damasio assumes that the neural systems underlying consciousness are hierarchically organized, with basic sensory and motor processing taking place at lower levels and more complex cognitive and emotional processing at higher levels. Consciousness arises when information from these different levels is coherently and uniformly integrated.

One of Damasio's key ideas is the concept of the body loop, which refers to the constant flow of information between the brain and the rest of the body. According to Damasio, this loop is essential for the generation of conscious experiences, as it allows the brain to constantly monitor and respond to the state of the body and the environment.

Damasio also emphasizes the importance of emotions in the generation of conscious experiences. He argues that emotions are not simply mental states, but complex physical reactions that play a crucial role in how we perceive the world and control our behavior.

Emotions and Feelings

In Damasio's theory, feelings and emotions are different but interconnected phenomena (Damasio, 2001). He defines emotions as direct signals from the body, indicating a positive or negative state. These complex patterns of physiological reactions are triggered by stimuli from the environment. In contrast, feelings are quasi second-order emotions, which correspond to the conscious perception of these physiological reactions.

Emotions are body states caused by so-called somatic markers, i.e., any-thing that any receptors inside the body can measure, such as blood pressure, heart rate, blood pH, blood sugar level, concentration of certain hormones, and so on and so forth. The totality of all somatic markers, i.e., all measure-ments from inside the body at a certain point in time, corresponds to the body state. This is represented and controlled by a multitude of brain regions, including the amygdala, the insula, and the prefrontal cortex. These brain regions work together to represent the state of the organism and to evoke and control the coordinated physiological reactions to environmental stimuli in a feedback loop, e.g., changes in heart rate, breathing, and hormone levels.

Feelings, on the other hand, arise from the conscious perception and interpretation of these physiological reactions. For example, the feeling of fear can be evoked by the physiological reactions associated with the emo-tion of fear, such as an increased heart rate and sweating. In Damasio's view, feelings give us a conscious representation of our bodily states and enable us to understand and respond to the world around us.

Damasio also emphasizes the role of emotions and feelings in deci-sion-making and behavior. He assumes that emotions provide us with important information about the environment and contribute to steer-ing our actions in an adaptive way. For example, the feeling of disgust can help us avoid potentially harmful substances, while the feeling of love can encourage us to seek social contacts.

Hierarchies of Consciousness

In Damasio's theory of consciousness, a hierarchy of neuronal and psycho-logical levels is assumed, which contribute to the generation of conscious experience. He distinguishes between proto-self, core-self, and extended or autobiographical self.

At the lowest level of this hierarchy is the *proto-self*, which arises from the constant monitoring of the internal body state and external stimuli. In humans and other mammals, it is continuously generated by neuronal activ-ity in the brainstem and thalamus, integrating sensory and motor signals from the entire body. The proto-self does not have the ability to recognize itself. It is a pure processing chain that reacts to inputs and stimuli like a machine, completely unconscious. According to this definition, every animal with a nervous system, from the snail to the dolphin, also has a proto-self.

The next level in the hierarchy is the *core-self*, which arises from the inte-gration of information from the proto-self with cognitive and emotional processing at a higher level. Here, higher common representations of feelings and objects from the environment are created, along with one's own actions and the emotional changes they cause. Thus, the core-self contributes to the experience of agency, self-responsibility, and self-awareness, and enables a more complex and differentiated understanding of ourselves as individuals. It can anticipate immediate reactions in its environment and adapt to them. It is also able to recognize itself and its parts in its own image of the world. This allows it to anticipate the world and react appropriately to it. The core-self is also fleeting and unable to linger for hours and forge complex plans. It is limited to the here and now. It remains constant throughout the organism's lifetime and is continuously recreated by mental patterns that arise from the interaction with internal body states and external stimuli (objects). These interactions lead to a representation referred to as the core-self, which is limited to the present moment. This can be imagined as a continuous stream of consciousness, which is constantly recreated by these interactions and gives our conscious experience a basic sense of coherence and continuity. Unlike many philosophical approaches, the core-self does not rely on representing the world in the form of language. Damasio even believes that the fixation on language has hindered progress in understanding the nature of consciousness. Damasio holds the view that every animal capable of adapting to its environment also has a core-self.

At the top of the hierarchy is the *extended* or *autobiographical self*, which encompasses our personal memories, beliefs, and narratives about ourselves and our lives by accessing extensive memory systems. It enables a human-like interaction with the environment and builds on the core-self. The ability to process language also falls into the category of the extended self and can be interpreted as a form of serialization of parallel consciousness contents. It is assumed that this level of self arises from the integration of information from the core-self with higher-level cognitive processes such as thinking and introspection. According to Damasio, the extended self is closely linked to our sense of identity and our understanding of ourselves as unique individuals with a past, present, and future. Higher mammals such as cats, dogs, chimpanzees, and dolphins also have an extended self, with this being most developed and thus most pronounced in humans.

While the proto-self must be hardwired from birth to ensure the survival of the organism, the other two levels of consciousness can be altered by experience. In particular, the extended self, which depends on various types of memory, develops gradually over the course of individual development. From birth, children have a proto- and core-self. The extended self, however, only comes into play during child development and becomes increasingly pronounced and differentiated. New functions are added through the progressive myelination and the associated functional integration of ever new cortex areas into the existing nervous system.

Explanatory Power of Damasio's Model

Due to its strongly mechanistic character, Damasio's model is able to explain some of the most important neurological consciousness disorders comparatively simply.

The vegetative state, also known as a coma, is a state in which a person appears to be awake but shows no conscious actions or reactions to their environment. People in a vegetative state usually have their eyes open and can maintain sleep-wake cycles, but are unable to respond to verbal, tactile, or painful stimuli. There are also no signs of conscious perception, thoughts, or feelings. The vegetative state can occur after a severe brain injury, stroke, or other brain disease. According to Damasio's model, the emergence of the vegetative state can be traced back to two different causes: The brain no longer receives information about the body's state, or the parts of the brain responsible for generating the core self are impaired.

The flow of sensory information from the body to the brain is crucial for the formation and maintenance of the core self. In a vegetative state, this flow can be interrupted or altered, which can prevent the brain from generating the core self. This interruption can have various causes, such as a traumatic brain injury, lack of oxygen in the brain, or another severe nervous system disease. Without the continuous feedback loop from the body's interior, the brain cannot build the core self and consciousness is lost. In some cases, the neural structures responsible for the formation of the core self can also be damaged, leading to a collapse of this basic level of consciousness. This can occur as a result of a traumatic brain injury or other neurological disorders. In both cases, the core self collapses—in one case because it lacks input (software problem), and in the other case because it can no longer process input (hardware problem). Since the core self is the basis for the autobiographical or extended self, this too is lost and consciousness disappears.

The Locked-in Syndrome is a condition in which a person is unable to move or speak due to damage to the brainstem, but is usually fully conscious. People with this syndrome can usually still move their eyelids and eye muscles to communicate. However, they are unable to control their muscles to actively move or speak, and are therefore proverbially trapped in their own body. Here too, we can try to interpret the symptoms in the context of Damasio's consciousness model.

Since both the flow of sensory information from the body to the brain and the neural structures responsible for generating the core self are not impaired in Locked-in Syndrome, the core self remains intact. Despite the almost complete lack of motor output, the brain still receives sensory input and can thus continue to generate the core self. This allows the person to maintain a basic level of self-perception and consciousness. However, Locked-in Syndrome interrupts the person's ability to interact with the world through motor actions, which can have long-term effects on the extended self (Kübler & Birbaumer, 2008), as the extended self is based on autobiographical memories, personal identity, and the ability to plan and think about the future. The inability to communicate and interact with the environment can lead to difficulties in maintaining and updating the extended self. However, the extended self does not completely disappear, as the individual is still able to process and create new memories based on sensory experiences.

Agnosia is a neurological phenomenon in which people have difficulty recognizing sensory information, even though the corresponding sensory organs are intact. There are a wide variety of different forms of agnosia in all sensory modalities. In visual agnosia, the recognition of visual stimuli is disturbed, while vision is intact. Examples are prosopagnosia, where sufferers can no longer recognize faces, object agnosia, which impairs object recognition, or color agnosia, where sufferers perceive the world in black and white only. Particularly bizarre is akinetopsia, which impairs motion perception. Patients suffering from it perceive the world only as a stroboscopic sequence of single images. Depending on the form, aphasias impair various aspects of language comprehension and speech. And there are many more of these impairments of the extended self.

Of course, in all these cases, the affected individuals are fully conscious. Their proto- and core self is completely intact. Only very specific functions of their extended self are isolatedly impaired.

Conclusion

The mind-body problem is far from solved and remains the subject of intense debates among philosophers, neuroscientists, and psychologists. However, it is clear that our understanding of the relationship between mind and body is essential for understanding the nature of human consciousness and subjective experience.

The debate between monism and dualism continues to shape philosophical discourse on the nature of the mind. While monism postulates a unified and interconnected worldview, dualism emphasizes the uniqueness of mind and body and suggests that there might be more behind our experiences than can be explained by physical phenomena alone.

Philosophy allows us to ask much broader questions than many other scientific disciplines (Laplane et al., 2019). It is capable of looking at the bigger picture and providing important insights into the relationships between different areas of knowledge. Philosophy is particularly important for the interdisciplinary efforts of cognitive science, where it helps to bridge gaps between different disciplines and pave new ways for research. Unlike scientific methods, philosophizing is a non-empirical approach that attempts to validate concepts through logical thinking and argumentation. Philosophers tend to ask questions rather than provide definitive answers, and their contributions often consist of challenging established assumptions and proposing new research approaches. However, for a more comprehensive understanding of the nature of consciousness, close collaboration between philosophy and neuroscience is required (Lamme, 2010). This means that while philosophy can provide valuable insights into theoretical concepts and broader ethical questions, it needs to be supplemented by empirical findings and experiments to reach a more comprehensive understanding.

The Integrated Information Theory (IIT) and the Global Workspace Theory (GWT) are the two most prominent neuroscientific theories attempting to explain the nature of consciousness. There are some similarities between these theories, but also some significant differences. One of the main differences between IIT and GWT is that they focus on different aspects of consciousness. IIT emphasizes the subjective experience of consciousness and attempts to quantify the amount of integrated information present in a particular system, while GWT emphasizes the cognitive and computational aspects of consciousness and the role of attention in shaping conscious experience. Another significant difference between the two theories is their approach to explaining the neural basis of consciousness. IIT assumes that consciousness arises from the integration of information in complex neural networks, while GWT assumes that consciousness arises from the global activation of a distributed network of brain regions that serve as a "workspace" for the integration and processing of information. In terms of the nature of conscious experience, IIT suggests that consciousness is a fundamental property of certain complex systems to influence themselves, while GWT assumes that conscious experience is a byproduct of cognitive processes that include attention and the global activation of neural networks.

Prominent representatives of both competing theories proposed an ambitious and large-scale experiment, distributed across many labs worldwide, in 2021 to decide between the two theories (Melloni et al., 2021). GWT predicts that consciousness arises from the global transmission and amplification of information through interconnected networks of prefrontal, parietal, and sensory cortex areas, while IIT suggests that consciousness is based on the intrinsic ability of a neural network to influence itself by generating maximally integrated information, with the posterior, parietal cortex being the ideal location for this. Various imaging techniques such as fMRI, EEG, and MEG, as well as sophisticated experimental paradigms, will now be used to test which prediction and thus which of the two theories is correct. We can look forward to the results. The author would not be surprised if the truth—as often in the history of science—lay somewhere in the middle and the researchers found activation in all mentioned cortex regions and thus evidence for both theories.

From the author's perspective, the processes described by both theories are probably necessary but not sufficient to actually generate consciousness. The only theory that offers a mechanistic explanation is Damasio's theory. While according to Damasio's framework the neural structures and processes of GWT and IIT are fully accounted for in the extended self or consciousness, his theory of somatic markers, the body loop, the proto-self and core self provide a basis on which the other higher-level processes can build. In the terminology of computer science, while GWT, IIT, and extended self correspond to a user interface with various "apps" (language, visual perception, etc.), the core self would be a kind of operating system and the proto-self a kind of BIOS.

References

Baars, B. J. (1994). A global workspace theory of conscious experience. In *Consciousness in Philosophy and Cognitive Neuroscience,* (pp. 149–171). Erlbaum.

Baars, B. J. (2017). The global workspace theory of consciousness. In *The Blackwell Companion to Consciousness,* (pp. 236–246). Wiley.

Baars, B. J., & Newman, J. (1994). A neurobiological interpretation of global workspace theory. *Consciousness in Philosophy and Cognitive Neuroscience,* 211–226.

Bayne, T., Seth, A. K., & Massimini, M. (2020). Are there islands of awareness? *Trends in Neurosciences, 43*(1), 6–16.

Boly, M., Massimini, M., Tsuchiya, N., Postle, B. R., Koch, C., & Tononi, G. (2017). Are the neural correlates of consciousness in the front or in the back of the cerebral cortex? Clinical and neuroimaging evidence. *Journal of Neuroscience,* *37*(40), 9603–9613.

Boly, M., Seth, A. K., Wilke, M., Ingmundson, P., Baars, B., Laureys, S., ... & Tsuchiya, N. (2013). Consciousness in humans and non-human animals: Recent advances and future directions. *Frontiers in Psychology, 4,* 625.

Casali, A. G., Gosseries, O., Rosanova, M., Boly, M., Sarasso, S., Casali, K. R., ... & Massimini, M. (2013). A theoretically based index of consciousness independent of sensory processing and behavior. *Science Translational Medicine, 5*(198), 198ra105–198ra105.

Casarotto, S., Comanducci, A., Rosanova, M., Sarasso, S., Fecchio, M., Napolitani, M., ... & Massimini, M. (2016). Stratification of unresponsive patients by an independently validated index of brain complexity. *Annals of Neurology, 80*(5), 718–729.

Chalmers, D. J. (1995). Facing up to the problem of consciousness. *Journal of Consciousness Studies, 2*(3), 200–219.

Cleeremans, A. (2005). Computational correlates of consciousness. *Progress in Brain Research, 150,* 81–98.

Cohen, M. A., & Dennett, D. C. (2011). Consciousness cannot be separated from function. *Trends in Cognitive Sciences, 15*(8), 358–364.

Crick, F., & Koch, C. (1990). Towards a neurobiological theory of consciousness. In *Seminars in the neurosciences* (Vol. 2, pp. 263–275). Saunders Scientific Publications.

Crick, F., & Koch, C. (2003). A framework for consciousness. *Nature Neuroscience, 6*(2), 119–126.

Damasio, A. R. (1999). *The feeling of what happens: Body and emotion in the making of consciousness.* Houghton Mifflin Harcourt.

Damasio, A. (2001). Fundamental feelings. *Nature, 413*(6858), 781–781.

Damasio, A., & Meyer, K. (2009). Consciousness: An overview of the phenomenon and of its possible neural basis. In S. Laureys & T. Giulio (Eds.), *The Neurology of Consciousness: Cognitive Neuroscience and Neuropathology* (Vol. 1, 10. Okt. 2008) (pp. 3–14). Academic.

Damasio, A. R. (2014). *Ich fühle, also bin ich: Die Entschlüsselung des Bewusstseins.* Ullstein eBooks.

Dehaene, S., & Changeux, J. P. (2004). Neural mechanisms for access to consciousness. *The Cognitive Neurosciences, 3,* 1145–1158.

Dehaene, S., Changeux, J. P., & Naccache, L. (2011). The Global Neuronal Workspace Model of Conscious Access: From Neuronal Architectures to Clinical Applications. In S. Dehaene & Y. Christen (Eds.), *Characterizing Consciousness: From Cognition to the Clinic?. Research and Perspectives in Neurosciences* Springer. https://doi.org/10.1007/978-3-642-18015-6_4.

Dehaene, S., Charles, L., King, J. R., & Marti, S. (2014). Toward a computational theory of conscious processing. *Current Opinion in Neurobiology, 25,* 76–84.

Dehaene, S., Kerszberg, M., & Changeux, J. P. (1998). A neuronal model of a global workspace in effortful cognitive tasks. *Proceedings of the National Academy of Sciences, 95*(24), 14529–14534.

Dehaene, S., Lau, H., & Kouider, S. (2017). What is consciousness, and could machines have it? *Science, 358*(6362), 486–492.

Dehaene, S., & Naccache, L. (2001). Towards a cognitive neuroscience of consciousness: Basic evidence and a workspace framework. *Cognition, 79*(1–2), 1–37.

Demertzi, A., Tagliazucchi, E., Dehaene, S., Deco, G., Barttfeld, P., Raimondo, F., … & Sitt, J. D. (2019). Human consciousness is supported by dynamic complex patterns of brain signal coordination. *Science Advances, 5*(2), eaat7603.

Dennett, D. C. (1991). *Consciousness explained.* Little, Brown and Company.

Graves, T. L., Maniscalco, B., & Lau, H. (2011). Volition and the function of consciousness. In *Conscious Will and Responsibility* (pp. 109–121) Graves.

Grossberg, S. (2017). Towards solving the hard problem of consciousness: The varieties of brain resonances and the conscious experiences that they support. *Neural Networks, 87*, 38–95.

Hameroff, S. (2001). Consciousness, the brain, and spacetime geometry. *Annals of the New York Academy of Sciences, 929*(1), 74–104.

Hameroff, S., Nip, A., Porter, M., & Tuszynski, J. (2002). Conduction pathways in microtubules, biological quantum computation, and consciousness. *Bio Systems, 64*(1–3), 149–168.

Hameroff, S., & Penrose, R. (1996a). Orchestrated reduction of quantum coherence in brain microtubules: A model for consciousness. *Mathematics and Computers in Simulation, 40*(3–4), 453–480.

Hameroff, S. R., & Penrose, R. (1996b). Conscious events as orchestrated spacetime selections. *Journal of Consciousness Studies, 3*(1), 36–53.

Koch, C. (2004). The quest for consciousness. *Engineering and Science, 67*(2), 28–34.

Koch, C. (2013). *Bewusstsein: Bekenntnisse eines Hirnforschers.* Springer.

Koch, C., & Hepp, K. (2006). Quantum mechanics in the brain. *Nature, 440*(7084), 611–611.

Koch, C., & Hepp, K. (2007). The relation between quantum mechanics and higher brain functions: Lessons from quantum computation and neurobiology. Citeseer. https://citeseerx.ist.psu.edu/viewdoc/download?doi=10.1.1.652.712& rep=rep1&type=pdf.

Koch, C., Massimini, M., Boly, M., & Tononi, G. (2016). Neural correlates of consciousness: Progress and problems. *Nature Reviews Neuroscience, 17*(5), 307–321.

Kriegeskorte, N., & Douglas, P. K. (2018). Cognitive computational neuroscience. *Nature Neuroscience, 21*(9), 1148–1160.

Kübler, A., & Birbaumer, N. (2008). Brain – computer interfaces and communication in paralysis: Extinction of goal directed thinking in completely paralysed patients? *Clinical Neurophysiology, 119*(11), 2658–2666.

Lamme, V. A. (2010). How neuroscience will change our view on consciousness. *Cognitive Neuroscience, 1*(3), 204–220.

Laplane, L., Mantovani, P., Adolphs, R., Chang, H., Mantovani, A., McFall-Ngai, M., ... & Pradeu, T. (2019). Why science needs philosophy. *Proceedings of the National Academy of Sciences, 116*(10), 3948–3952.

Mashour, G. A., Roelfsema, P., Changeux, J. P., & Dehaene, S. (2020). Conscious processing and the global neuronal workspace hypothesis. *Neuron, 105*(5), 776–798.

Melloni, L., Mudrik, L., Pitts, M., & Koch, C. (2021). Making the hard problem of consciousness easier. *Science, 372*(6545), 911–912.

Nagel, T. (1974). What is it like to be a bat? *The Philosophical Review, 83*(4), 435–450.

Newman, J. B., & Baars, B. J. (1993). A neural attentional model for access to consciousness: A global workspace perspective. In *Concepts in Neuroscience, 4*(2), 109–121. Graves.

Nielsen, M., & Chuang, I. (2002). *Quantum computation and quantum information*. Cambridge University Press.

Penrose, R. (1989). *The emperor's new mind*. Oxford University Press.

Penrose, R. (1994). Mechanisms, microtubules and the mind. *Journal of Consciousness Studies, 1*(2), 241–249.

Rees, G., Kreiman, G., & Koch, C. (2002). Neural correlates of consciousness in humans. *Nature Reviews Neuroscience, 3*(4), 261–270.

Reggia, J. A., Katz, G., & Huang, D. W. (2016). What are the computational correlates of consciousness? *Biologically Inspired Cognitive Architectures, 17*, 101–113.

Sergent, C., & Dehaene, S. (2004). Neural processes underlying conscious perception: Experimental findings and a global neuronal workspace framework. *Journal of Physiology-Paris, 98*(4–6), 374–384.

Seth, A. (2009). Explanatory correlates of consciousness: Theoretical and computational challenges. *Cognitive Computation, 1*(1), 50–63.

Seth, A. (2021). *Being you: A new science of consciousness*. Penguin.

Seth, A. K., & Bayne, T. (2022). Theories of consciousness. *Nature Reviews Neuroscience, 23*(7), 439–452.

Seth, A. K., Dienes, Z., Cleeremans, A., Overgaard, M., & Pessoa, L. (2008). Measuring consciousness: Relating behavioural and neurophysiological approaches. *Trends in Cognitive Sciences, 12*(8), 314–321.

Tegmark, M. (2000). Importance of quantum decoherence in brain processes. *Physical Review E, 61*(4), 4194.

Tegmark, M. (2014). Consciousness is a state of matter, like a solid or gas. *New Scientist, 222*(2964), 28–31.

Tononi, G. (2004). An information integration theory of consciousness. *BMC Neuroscience, 5*, 1–22.

Tononi, G., Boly, M., Massimini, M., & Koch, C. (2016). Integrated information theory: From consciousness to its physical substrate. *Nature Reviews Neuroscience,* *17*(7), 450–461.

Tononi, G., & Koch, C. (2008). The neural correlates of consciousness: An update. *Annals of the New York Academy of Sciences, 1124*(1), 239–261.

10

Free Will

Life is like a card game. The hand you receive is deterministic;
how you play it is free will.

Jawaharlal Nehru

Is Free will Just a Pious Wish?

The question about the nature of consciousness is closely linked to the question of the existence of free will, that is, the ability to choose freely between different possible courses of action, free from external influences or desires (O'Connor, 1972; Kane, 2001; Watson, 2003; Harris, 2012; Ekstrom, 2018). The concept of free will is closely linked to notions of moral responsibility, praise, guilt, sin, and other ethical and legal concepts (Roth, 2004, 2016; Roth et al., 2006; Lampe et al., 2008).

In principle, free will also means the ability to make decisions that are not determined by past events or so-called deterministic causal chains, i.e., that the actor can choose between different possible outcomes and that the result of his decision is not predetermined by previous events (James, 1884; Van Inwagen, 1975). In classical determinism, however, it is assumed that all events are determined by past causes and thus only one course is possible (Earman, 1986).

P. Krauss, *Artificial Intelligence and Brain Research*,
https://doi.org/10.1007/978-3-662-68980-6_10

The Laplace's Demon, named after the French mathematician Pierre-Simon Laplace, is a philosophical and physical thought experiment that examines the concept of determinism (Van Strien, 2014; Kožnjak, 2015). The thought experiment involves a hypothetical demon who knows the position and speed of all particles in the universe at a certain point in time. With this knowledge, the demon would be able to predict the future of the universe with absolute certainty, as he would know the exact outcome of each event based on the physical laws. The Laplace's Demon is often used to illustrate the idea that if determinism is true, the future of the universe is already predetermined and all events are necessary and inevitable. This notion has significant implications for the concept of free will, because if the future is predetermined, there seems to be no room for real decisions or options for action. Incompatibilists therefore argue that if determinism is true, free will is not possible.

However, Laplace's Demon is also the subject of various criticisms and objections. One objection is that the demon's complete knowledge of the universe could be impossible, as Heisenberg's uncertainty principle implies that there are limits to our ability to simultaneously measure the position and speed of particles with arbitrary accuracy (Robertson, 1929; Busch et al., 2007). Moreover, it is argued that even if the demon's knowledge were possible, it would not necessarily imply complete predictability, as quantum physics gives the universe an inherent randomness and unpredictability (Dirac, 1925, 1926; Schrödinger, 1926).

And finally, determinism and predictability are not necessarily identical (Van Kampen, 1991; Loewer, 2001), as anyone who has ever been annoyed by the wrong weather forecast knows. The dynamics of complex systems are sensitive to initial conditions, which in principle would have to be known with infinite accuracy in order to predict the further course exactly (Strogatz, 2018). Otherwise, the slightest deviations over time can lead to exponentially growing differences in behavior (Murphy, 2010). This is the infamous butterfly effect (Lorenz, 2000). The weather is certainly a deterministic system and follows the causal laws of cause and effect, but it is far from being exactly predictable into the distant future (Koch, 2009).

Some philosophers therefore advocate a compatibilist view of free will, which states that free will is compatible with determinism (Vihvelin, 2013). Although the universe is largely deterministic, there is still enough randomness and chaos in the world to save free will. Others argue that the apparent contradiction between free will and determinism is a category mistake. The opposite of determinism is not free will, but randomness and chaos. And the opposite of free will is not determinism, but compulsion (Mittelstraß et al., 2004; Hallett, 2009; 2011) (Fig. 10.1).

Fig. 10.1 Free will. The existence of free will is disputed in philosophy and neuroscience. However, there is no doubt about the existence of the village of the same name in the district of Schleswig-Flensburg in the north of Germany.

The Libet Experiment

The Libet Experiment, named after Benjamin Libet, is a famous and controversial study in neuroscience and psychology, intended to investigate the nature of free will and the timing of conscious decisions (Libet et al., 1983; Libet, 1985). The experiment was conducted in the early 1980s to determine whether the conscious decision to perform a voluntary action occurs before or after the brain triggers the action.

The basic experimental setup consisted of four components. First, volunteers were asked to perform simple voluntary movements, such as bending the wrist or pressing a button. Meanwhile, their brain activity was monitored using electroencephalography (EEG), focusing on the readiness potential. This refers to the slow increase in electrical activity in the brain that precedes voluntary movements and can thus be interpreted as an indication of the emergence of a conscious intention. Second, participants were shown a clock-like device that served as a timer. A rotating dot allowed them to report the exact time of their conscious decision to perform the voluntary movement. Third, the participants were asked to indicate the position of the rotating dot on the clock at the time of their conscious decision. And finally,

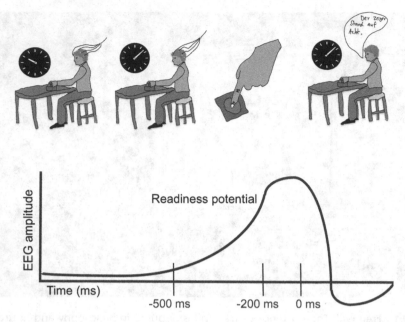

Fig. 10.2 The Libet Experiment. A subject performs a voluntary movement while his brain activity is monitored using EEG. A clock-like device allows him to record the time of his conscious decision. He then indicates the position of the rotating pointer on the clock at the decision time. Finally, brainwave measurements and decision reports are compared. Results suggest that unconscious processes already initiate actions (−500 ms) before the conscious decision to do so is made (−200 ms) and the action is then finally executed (0 ms)

fourth, the results of the brainwave measurement were compared with the decision reports to determine whether the participants' conscious decision was indeed the cause of the voluntary movement (Fig. 10.2).

The most important result of Libet's experiment was that the readiness potential in the brain already increased significantly (500−300 milliseconds) before the time when the participants reported that they had become aware of their decision to perform the movement. Libet concluded from this that unconscious processes in the brain initiate voluntary actions before conscious decisions are made.

Conclusion

The results of the Libet experiment seem to argue against the existence of free will. They have therefore been widely discussed and have led to heated controversies. Critics claim that the methodology of the experiment is flawed and the conclusions about free will are too deterministic. It is also argued that the decision to perform a certain movement at a certain time oversimplifies the nature of free will. For example, this simple action is not comparable to freely deciding for or against a certain course of study, a certain place of residence, or a life partner.

From the author's perspective, it would be highly strange if our (free) decisions were completely independent of our genetic, neural, and autobiographical history. Of course, important life decisions like the ones mentioned above are made against the backdrop of our personal experience and life history, which in turn are stored in the brain as knowledge and memories (Hallet, 2007; Lau, 2009).

Imagine if Libet's results had been exactly the opposite: The test subjects report their decision to perform the movement before the corresponding neural activity is measurable. This result would have been much more disturbing, as it would have implied that there are biology-independent mental processes that are capable of causing changes in the brain. This would have been de facto evidence of dualistic views on the relationship between brain and mind.

Some even went so far as to question the practice of punishment with the argument that if the will is not free, the person cannot be held responsible for their actions (Roth, 2010, 2012; Singer, 2020). This of course overlooks that punishments can still be meaningful as they serve different purposes and work on different levels (Feinberg, 1965; Van den Haag, 1975; Jakobs, 2011; Kühl, 2017). First, victims of crimes, even if the existence of free will is questioned, still have a need for justice, atonement, and reparation. Second, resocialization, regardless of the question of free will, can contribute to rehabilitating the offender and imparting the skills and knowledge to become a constructive member of society. In this context, resocialization focuses on changing behaviors and thought patterns that have led to crimes, not on whether the offender consciously made these decisions. Third, the punishment of offenders can serve as a deterrent for potential imitators by clarifying the consequences of crimes. Even if people do not have free will, they can still respond to incentives and deterrence. The sanction system can thus help to reduce the frequency of crimes by deterring potential offenders from criminal behavior. And finally, the imprisonment of an offender can protect society from further crimes by this person. Even if the offender does not have free will, he could still pose a danger to society if left free. Thus, punishment and imprisonment of an offender can remain meaningful and justified even if the existence of free will is questioned (Viney, 1982; Stroessner & Green, 1990; Hallett, 2007; Mobbs et al., 2009; Hodgson, 2009; Vincent et al., 2011; Focquaert et al., 2013).

In addition to the impacts on human decision-making, the Libet experiments also raise interesting questions about artificial intelligence. Although AI is fundamentally deterministic, these experiments can influence our understanding of AI decisions, responsibility, and moral action. Furthermore, AI can serve as a tool to explore and understand the concept of free will in humans.

References

Busch, P., Heinonen, T., & Lahti, P. (2007). Heisenberg's uncertainty principle. *Physics Reports, 452*(6), 155–176.

Dirac, P. A. M. (1925). The fundamental equations of quantum mechanics. Proceedings of the Royal Society of London. *Series A, Containing Papers of a Mathematical and Physical Character, 109*(752), 642–653.

Dirac, P. A. M. (1926). On the theory of quantum mechanics. Proceedings of the Royal Society of London. *Series A, Containing Papers of a Mathematical and Physical Character, 112*(762), 661–677.

Earman, J. (1986). *A primer on determinism* (Vol. 37). Springer Science & Business Media.

Ekstrom, L. (2018). *Free will.* Routledge.

Feinberg, J. (1965). The expressive function of punishment. *The Monist, 49*(3), 397–423.

Focquaert, F., Glenn, A. L., & Raine, A. (2013). Free will, responsibility, and the punishment of criminals. *The future of punishment,* 247–274.

Hallett, M. (2007). Volitional control of movement: The physiology of free will. *Clinical Neurophysiology, 118*(6), 1179–1192.

Hallett, M. (2009). Physiology of volition. *Downward causation and the neurobiology of free will,* 127–143.

Hallett, M. (2011). Volition: How physiology speaks to the issue of responsibility. *Conscious will and responsibility,* 61–69.

Harris, S. (2012). *Free will.* Simon and Schuster.

Hodgson, D. (2009). Criminal responsibility, free will, and neuroscience. *Downward causation and the neurobiology of free will,* 227–241.

Jakobs, G. (2011). *Strafrecht: Allgemeiner Teil.* de Gruyter.

James, W. (1884). *The dilemma of determinism* (pp. 1878–1899). Kessinger Publishing.

Kane, R. (Hrsg.). (2001). *Free will.* Wiley.

Koch, C. (2009). Free will, physics, biology, and the brain. *Downward causation and the neurobiology of free will,* 31–52.

Kožnjak, B. (2015). Who let the demon out? Laplace and Boscovich on determinism. *Studies in History and Philosophy of Science Part A, 51,* 42–52.

Kühl, K. (2017). *Strafrecht: Allgemeiner Teil.* Vahlen.

Lampe, E. J., Pauen, M., Roth, G., & Verlag, S. (Eds.). (2008). *Willensfreiheit und rechtliche Ordnung.* Suhrkamp.

Lau, H. C. (2009). Volition and the function of consciousness. *Downward causation and the neurobiology of free will,* 153–169.

Libet, B., Wright, E. W., Jr., & Gleason, C. A. (1983). Preparation-or intention-to-act, in relation to pre-event potentials recorded at the vertex. *Electroencephalography and Clinical Neurophysiology, 56*(4), 367–372.

Libet, B. (1985). Unconscious cerebral initiative and the role of conscious will in voluntary action. *Behavioral and brain sciences, 8*(4), 529–539.

Loewer, B. (2001). Determinism and chance. *Studies in History and Philosophy of Science Part B: Studies in History and Philosophy of Modern Physics, 32*(4), 609–620.

Lorenz, E. (2000). The butterfly effect. *World Scientific Series on Nonlinear Science Series A, 39,* 91–94.

Mittelstraß, J., Menzel, R., Singer, W., & Nida-Rümelin, J. Zur Freiheit des Willens II: Streitgespräch in der Wissenschaftlichen Sitzung der Versammlung der Berlin-Brandenburgischen Akademie der Wissenschaften am 2. Juli 2004.

Mobbs, D., Lau, H. C., Jones, O. D., & Frith, C. D. (2009). *Law, responsibility, and the brain* (pp. 243–260). Springer.

Murphy, R. P. (2010). *Chaos theory.* Ludwig von Mises Institute.

O'Connor, D. J. (1972). *Free will.* Springer.

Robertson, H. P. (1929). The uncertainty principle. *Physical Review, 34*(1), 163.

Roth, G. (2004). Freier Wille, Verantwortlichkeit und Schuld. *Zur Freiheit des Willens,* 63–70.

Roth, G., Lück, M., & Strüber, D. (2006). „Freier Wille" und Schuld von Gewaltstraftätern aus Sicht der Hirnforschung und Neuropsychologie. *Neue Kriminalpolitik, 18*(2), 55–59.

Roth, G. (2010). Free will: Insights from neurobiology. *Homo novus—a human without illusions,* 231–245.

Roth, G. (2012). Über objektive und subjektive Willensfreiheit. *Theory of Mind: Neurobiologie und Psychologie sozialen Verhaltens,* 213–223.

Roth, G. (2016). Schuld und Verantwortung: Die Perspektive der Hirnforschung. *Biologie in unserer Zeit, 46*(3), 177–183.

Schrödinger, E. (1926). SCHRÖDINGER 1926C. *Annalen der Physik, 79,* 734.

Singer, W. (2020). *Ein neues Menschenbild?: Gespräche über Hirnforschung.* Suhrkamp.

Strogatz, S. H. (2018). *Nonlinear dynamics and chaos with student solutions manual: With applications to physics, biology, chemistry, and engineering.* CRC Press.

Stroessner, S. J., & Green, C. W. (1990). Effects of belief in free will or determinism on attitudes toward punishment and locus of control. *The Journal of Social Psychology, 130*(6), 789–799.

Van den Haag, E. (1975). *Punishing criminals* (Vol. 10). Basic Books.

Van Inwagen, P. (1975). The incompatibility of free will and determinism. *Philosophical Studies, 27*(3), 185–199.

Van Kampen, N. G. (1991). Determinism and predictability. *Synthese, 89,* 273–281.

Van Strien, M. (2014). On the origins and foundations of Laplacian determinism. *Studies in History and Philosophy of Science Part A, 45,* 24–31.

Vihvelin, K. (2013). *Causes, laws, and free will: Why determinism doesn't matter.* Oxford University Press.

Vincent, N. A., Van de Poel, I., & Van Den Hoven, J. (Eds.). (2011). *Moral responsibility: Beyond free will and determinism* (Vol. 27). Springer Science & Business Media.

Viney, W., Waldman, D. A., & Barchilon, J. (1982). Attitudes toward punishment in relation to beliefs in free will and determinism. *Human Relations, 35*(11), 939–949.

Watson, G. (Eds.). (2003). *Free will.* Oxford readings in philosophy.

Part II

Artificial Intelligence

In the second part of the book, the focus will be on the most important ideas and concepts of Machine Learning and Artificial Intelligence. Since mathematical details can be as daunting as neuroanatomical and molecular biological ones, this part will also avoid an in-depth mathematical presentation. The focus is on a general understanding and an overview of the big picture and the most important ideas and concepts, in order to grasp the essence and potentials of AI and Machine Learning. This approach helps us to build a bridge between the two disciplines of AI and brain research and to work out their similarities and differences without getting lost in technical details.

11

What is Artificial Intelligence?

Despite all the hype and excitement about AI, it is still very limited compared to human intelligence.

Andrew Ng

History of AI

The idea that human intelligence or the cognitive processes that humans use could be mechanized or automated is very old. One of the first mentions of this idea can be found in Julien Offray de La Mettrie's work *L'Homme Machine,* published in 1748. Another theoretical precursor of AI is the Laplace's Demon, which we have already encountered in the chapter on free will. It is named after the French mathematician, physicist, and astronomer Pierre-Simon Laplace. The concept is based on the idea that the entire universe functions like a mechanical machine, similar to a clock, and that the human mind and intelligence function in the same way.

Artificial Intelligence (AI) is based on the belief that human thinking can be structured and systematized. The roots of this concept go back to the 1st millennium BC, when Chinese, Indian, and ancient Greek philosophers developed structured techniques for formal reasoning. Later, in the seventeenth century, philosophers like René Descartes and Gottfried Wilhelm Leibniz tried to formalize rational thinking and make it as precise as algebra or geometry. They considered thinking as equivalent to the manipulation

P. Krauss, *Artificial Intelligence and Brain Research*,
https://doi.org/10.1007/978-3-662-68980-6_11

of symbols. This model later served as the basis for research in the field of Artificial Intelligence.

The study of mathematical logic played a crucial role in the development of AI in the twentieth century. One of the most important contributions came from George Boole and Gottlob Frege, who laid the foundations for the formal manipulation of symbols. These works established a set of rules and principles for the manipulation of symbols in a logical system, which proved essential for the development of intelligent machines.

Another important contribution was the Church-Turing thesis, a fundamental concept in computer science. It states that any mathematical problem that a human can solve with paper and pencil can also be solved by a machine, specifically a mechanical machine that can process zeros and ones. This concept laid the foundation for the development of digital computers and the theory of computing.

The Turing Machine, a theoretical computing model proposed by Alan Turing, was a crucial breakthrough in the development of Artificial Intelligence. It is a simple and abstract model that captures the essential features of any mechanical machine capable of performing abstract symbol manipulations. The Turing Machine allowed researchers to formalize the concept of computing and develop algorithms that can be executed by machines.

The Dartmouth Conference, which took place in the summer of 1956 at Dartmouth College in Hanover, New Hampshire, is considered the founding event of the academic field of Artificial Intelligence (AI). The conference was a six-week workshop titled "Dartmouth Summer Research Project on Artificial Intelligence," organized by John McCarthy as part of a research project funded by the Rockefeller Foundation. The term "Artificial Intelligence" was used for the first time in the announcement of this conference. Some of the brightest minds in computer science and related fields attended the conference, including McCarthy himself, Marvin Minsky, Nathaniel Rochester, and Claude Elwood Shannon, the founder of information theory. During the conference, the participants discussed the possibility of creating machines that can think and argue like humans. They examined various AI approaches such as logic-based systems, neural networks, and heuristic search algorithms. They also discussed the ethical implications of developing intelligent machines and the question of whether it is even possible to develop truly intelligent machines. The Dartmouth Conference established AI as an academic discipline and provided a roadmap for future research in this field.

There is a close connection between the research direction of Artificial Life and AI. The long-term goal of AI is to develop an intelligent agent capable of understanding or learning any intellectual task that can be accomplished by a human or other living being. This is also referred to as strong AI or Artificial General Intelligence *(Artificial General Intelligence, AGI).*

Concept Clarification

Artificial Intelligence generally refers to machines capable of intelligent behavior (Russell, 2010). Unlike the natural intelligence of animals and humans, which is based on biological processes, AI is based on algorithms and software developed by humans. AI systems are programmed to perform tasks that normally require human intelligence, such as pattern recognition, learning from experience, decision-making, and problem-solving. An important aspect of AI systems is their ability to be autonomous and adaptable. They can learn from experiences and feedback and improve over time. This process is also referred to as Machine Learning and includes various techniques such as supervised learning, unsupervised learning and reinforcement learning. There are many different types of AI systems, each specialized in their own way. Natural language processing, for example, allows machines to understand and respond to human language (Görz & Schneeberger, 2010; Russell, 2010).

Expert systems use knowledge in the form of symbols and rules to solve problems and make decisions. Autonomous robots are another example of AI systems specialized in physical actions. They can perform tasks such as navigation, object manipulation, and interaction with the environment. Multi-agent systems, on the other hand, are a type of swarm of autonomous agents that work together to achieve a common goal (Görz & Schneeberger, 2010; Russell, 2010).

When artificial intelligence is discussed in the media and politics today, it usually refers to only a relatively small part of these approaches. All the spectacular successes and breakthroughs of the last one or two decades—whether it's composing music, playing Go, generating text, or classifying images—are based on machine learning and pattern recognition, a subfield of AI, and here in particular on deep learning, which is only a part of machine learning. In public discourse, the terms artificial intelligence, machine learning, and deep learning are often used synonymously today (Fig. 11.1). Often in the context of AI, "the algorithms" are mentioned. An algorithm is similar to

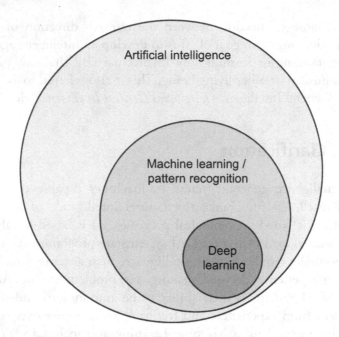

Fig. 11.1 Subareas of Artificial Intelligence. The terms Artificial Intelligence, Machine Learning, and Deep Learning are often seen as interchangeable today and are used synonymously in public discourse, although they actually stand in a hierarchical relationship to each other

a cooking recipe, a step-by-step guide to solving a problem or performing a specific task. It consists of an ordered sequence of instructions formulated in such a way that they can be executed by a machine, a computer, or a human.

Machine Learning and Pattern Recognition

Machine Learning deals with the development of algorithms and statistical models that enable computers to learn from data and make predictions or decisions without being explicitly programmed (Bishop & Nasrabadi, 2006). The focus is on analyzing large amounts of data to recognize patterns and draw conclusions. Machine learning is an important part of modern data analysis and is used in many areas, including image and speech recognition, autonomous driving, and personalized advertising. An important concept of machine learning is pattern recognition. Patterns in data are recognized to make predictions or decisions. To do this, algorithms and models must be developed that can learn from data and recognize patterns.

The human brain also uses pattern recognition to process and interpret sensory impressions. It has specialized regions responsible for certain types of pattern recognition. The brain's ability to learn patterns and adapt to new patterns is crucial for intelligent, goal-directed behavior and our ability to interact with the world.

Deep Learning

Deep Learning is a branch of Machine Learning that focuses on artificial neural networks with multiple layers of interconnected neurons. The more layers a neural network has, the "deeper" it is. Modern architectures can have hundreds of layers and are capable of processing and learning from large amounts of data by recognizing complex patterns and establishing abstract connections. One of the greatest achievements of Deep Learning is the ability to train neural networks to process large amounts of unstructured data, such as images or voice recordings, and recognize complex correlations and patterns. This has led to breakthroughs in many application areas, including language processing, image recognition and segmentation, and has enabled advances in medical research, natural sciences, and other fields (Schmidhuber, 2015; LeCun et al., 2015; Goodfellow et al., 2016).

The main advantage of Deep Learning is the ability to recognize and learn meaningful patterns in complex and unstructured data without the need for human expertise. Deep Learning models can even work with data that is difficult or impossible for the human mind to interpret, such as digital X-rays. This makes Deep Learning an enormously powerful tool in many fields.

Despite the incredible advances in recent years, there are still some challenges in the development and use of Deep Learning models. One of the biggest problems is the difficulty of interpreting and explaining the workings of deep neural networks. It is often hard to understand why a particular model made a certain decision or how it recognized certain patterns.

Conclusion

The successes of AI have been remarkable in recent years and cover a wide range of applications. One of the most impressive advances is the ability of AI systems to understand human language and process natural language. This has led to significant advances in speech recognition and translation. AI has also made progress in image processing by being able to recognize and analyze images. This has contributed to the development of face recognition, object detection, and medical imaging. Many practical applications of Artificial Intelligence (AI) are often integrated so quickly into everyday products that they are no longer perceived as AI. An example of this is text recognition, which is now standard in many smartphones and is taken for granted.

Because of this phenomenon known as the AI effect, it may seem that AI research is only struggling with difficult problems that it has not yet solved, such as the ability to make complex decisions in dynamic environments, or the problem that it is difficult to interpret and explain the workings of deep neural networks.

A famous quote by computer scientist Larry Tesler, "Intelligence is what machines have not yet done," expresses this.

References

Bishop, C. M., & Nasrabadi, N. M. (2006). *Pattern recognition and machine learning* (Vol. 4, No. 4, p. 738). Springer.

Goodfellow, I., Bengio, Y., & Courville, A. (2016). *Deep learning*. MIT press.

Görz, G., & Schneeberger, J. (Eds.). (2010). *Handbuch der künstlichen Intelligenz*. De Gruyter.

LeCun, Y., Bengio, Y., & Hinton, G. (2015). *Deep learning. nature, 521*(7553), 436–444.

Russell, S. J. (2010). *Artificial intelligence a modern approach*. Pearson Education, Inc.

Schmidhuber, J. (2015). Deep learning in neural networks: An overview. *Neural networks, 61*, 85–117.

12

How Does Artificial Intelligence Learn?

After my training as a computer scientist in the 90s, everyone knew that AI did not work. People tried. They tried with neural networks, but nothing worked.

Sergey Brin

Artificial Neuron

As in the brain, the neuron is also the fundamental processing unit in many areas of AI. Artificial neurons are simplified mathematical models of their biological counterparts. They receive their input in the form of real numbers through several input channels. As in the brain, this input can either come "from outside" or from other neurons. Each input channel has a weight, which corresponds to the synapse in biology, with which the respective input is multiplied. Subsequently, all weighted inputs of the neuron are summed up. Since all inputs of a neuron can be summarized into an input vector and analogously all corresponding weights can be summarized into a weight vector, the weighted sum of the inputs corresponds to the scalar product of the input and weight vector. This then goes into an activation function to determine the output of the neuron. In the McCulloch-Pitts neuron (McCulloch & Pitts, 1943), which represents the earliest and simplest artificial neuron, the activation function corresponds to a threshold function, i.e., it is compared whether the weighted sum of the inputs is greater or smaller than a certain threshold. If it is greater, the neuron sends an output (a one), otherwise not (which corresponds to a zero). Thus, the

P. Krauss, *Artificial Intelligence and Brain Research*, https://doi.org/10.1007/978-3-662-68980-6_12

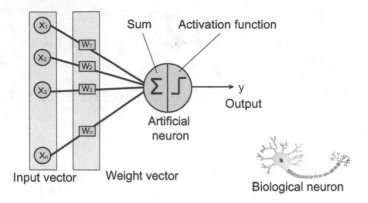

Fig. 12.1 Artificial Neuron. The basic processing unit in Artificial Intelligence. Analogous to its biological model, it takes input x through several channels and multiplies it with associated weights w, similar to the synapses in the brain. These weighted inputs are summed up and processed through an activation function to determine the output y of the neuron. In simple models like the McCulloch-Pitts neuron, the activation function is a threshold function, leading to binary outputs (0 or 1). More complex models use advanced activation functions for continuous outputs

perceptron is a binary classifier. However, more complicated activation functions are usually used, which lead to continuous outputs (Schmidhuber, 2015; LeCun et al., 2015; Goodfellow et al., 2016) (Fig. 12.1).

Artificial Neural Networks

From these artificial neurons, artificial neural networks can now be constructed by connecting several of them together, with a weight for each connection. Each neuron receives input signals from other neurons or from external sources, then processes this input and calculates an output signal, which is then transmitted further to other neurons or as output to the outside.

Depending on the architecture of the network, the neurons can be arranged in different so-called layers, with several neurons in one layer. As a rule, all neurons of a layer receive their input from the same source and send their output to the same target. In a neural network, one distinguishes between input layer, output layer and (usually several) intermediate layers or also hidden layers.

The weights of a neural network are summarized in the weight matrix. The columns of the weight matrix correspond to the weight vectors of the neurons. This matrix can either contain all pairwise weights of all neurons of the network including all self-connections (Autapses). In this case, one speaks of a complete weight matrix. However, a weight matrix can also only contain the forward-directed weights between the neurons of two successive layers of a network.

In the simplest case, the network consists only of an input and an output layer. These networks, also known as Perceptron (Rosenblatt, 1958), calculate a weighted sum of the inputs and then apply an activation function to generate the output. Due to this simple architecture, these neural networks can only learn linear classifications, i.e., the data in the input space must be linearly separable (Figs. 12.2 and 12.3).

Due to their linear character, two-layer networks are only limitedly able to solve classification tasks. A famous example that illustrates the limits of two-layer neural networks is the XOR problem. The XOR function (eXclusive OR) is a binary operation that outputs 1 when the two input values are different and 0 when they are the same. It turns out that this problem is not linearly solvable and thus cannot be learned by two-layer networks.

To overcome this limitation and solve problems with more complex decision boundaries, multi-layer neural networks with one or more hidden layers can be used. These hidden layers allow the network to learn nonlinear transformations of the input data, so they can capture more complex relationships and decision boundaries, making them ultimately more versatile and powerful for various tasks.

In fact, the so-called universal approximation theorem states that in principle, a neural network with a single hidden layer and a finite number of

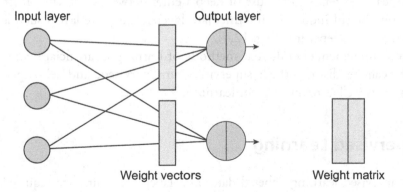

Fig. 12.2 Two-layer network. The perceptron consists only of an input and an output layer. The weight vectors are summarized as a weight matrix

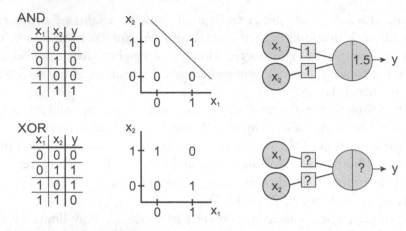

Fig. 12.3 The XOR problem. A two-layer neural network cannot solve the XOR problem. There is no suitable set of weights and threshold, as the two solutions (0, 1) are not linearly separable, i.e., the decision boundary between the classes is not a straight line. However, the logical AND can be solved with a simple two-layer network, as the solutions here are linearly separable

neurons can approximate any continuous function with arbitrary accuracy (Cybenko, 1989). However, usually neither the number of neurons nor the specific architecture of the network is clear. The theorem only guarantees the existence of such a network. Moreover, the theorem does not state that a network with a single hidden layer is always the best choice for a particular approximation problem or that it converges quickly during training.

In practice, therefore, neural networks with multiple hidden layers are often used to approximate complex functions, as they often achieve higher accuracy with fewer neurons than a network with a single hidden layer. Deep learning refers to the use of deep neural networks, which consist of a large number of hidden or intermediate layers. The more layers such a network has, the deeper it is (Fig. 12.4).

Four fundamentally different methods of learning in artificial neural networks can be distinguished: supervised, unsupervised, and self-supervised learning as well as reinforcement learning.

Supervised Learning

For supervised learning, labeled data, i.e., label-data pairs, are required. An application example is image classification. Here, there must be a label for each image with information about what is seen in the image or to which

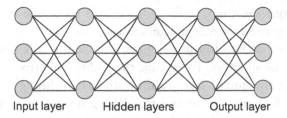

Input layer   Hidden layers   Output layer

Fig. 12.4 Multi-layer neural network. Artificial neural networks are complex structures built from artificial neurons and interconnected. In such a structure, several neurons can be present in one layer, connected to other neurons via weights. Hidden layers or intermediate layers are located between the input and the output layer. Deep learning refers to the use of deep neural networks, which consist of a large number of hidden layers. The more layers a neural network has, the deeper it is. All these networks are feedforward architectures, i.e., forward-directed networks, as the input flows in one direction without feedback loops, namely from the input to the output layer

category the image belongs. In an image dataset, "cat", "apple" or "car" can be possible labels. The labels correspond to the desired output of the model in supervised learning.

For the training of the neural network, the images are first broken down into individual pixels, with the color or grayscale values coded as numerical values serving as input channels for the first layer of the neural network. In a fully connected network, each input channel is connected to each neuron of the next layer. Activation is achieved by applying the weights to the inputs. This process can be repeated as often as desired in the various layers. At the end of the neural network is the output layer, which makes a decision, e.g., the classification of the image. If the network is already trained and has learned to classify a certain image correctly, the task is completed. However, if the network is not yet sufficiently trained, an error may occur. For example, in the output layer, the "apple" neuron could be most strongly activated, even though the input layer received an image of a "banana" as input.

During training, the desired output is compared with the actual output and an error is calculated using the loss or cost function. Since it is similar to a teacher monitoring the performance of the network and returning corresponding error messages, this learning paradigm is referred to as supervised learning. The desired output for a given input, such as the appropriate label (apple, banana) for an image, is usually provided by humans. You are probably familiar with these CAPTCHAs (completely automated public Turing test to tell computers and humans apart) on certain websites, where you are asked to confirm that you are not a robot. For example, you are asked to click on all images showing a traffic light. In this way, large amounts of labeled data for supervised learning are generated.

The goal of supervised learning is to minimize the cost function, i.e., to reduce the sum of all errors. To do this, the individual errors are propagated backwards from the output layer to the input layer through all layers. This process is therefore referred to as Error-Backpropagation and allows for a readjustment of the synaptic weights. This means that the errors are used to change the weights between two layers in such a way that the correct output is produced with a slightly higher probability the next time. This is done by calculating so-called gradients, which indicate the directions in which the weights need to be changed. This basic technique of machine learning is therefore also referred to as Gradient Descent (Schmidhuber, 2015; LeCun et al., 2015; Goodfellow et al., 2016).

In a figurative sense, the error landscape corresponds to a mountain range with many peaks and valleys. The higher you are, the greater the total error of the network. The gradients indicate at each point the direction in which it goes down steepest. By taking a small step in the direction of the steepest descent at each time step, you eventually end up in a valley, which corresponds to a lower total error of the network.

At the end of the training, the test accuracy of the neural network is usually determined. This is done by calculating the ratio of correctly predicted or classified objects to the total number of objects in the dataset. For example, if a model trained to classify images correctly classifies 90 out of 100 images, then the test accuracy of the model is 90%.

Usually, the entire available dataset is randomly divided into a training and a test dataset, usually in a ratio of 80 to 20, a practice referred to as dataset splitting. The idea behind this is that you want to test how well the neural network can generalize, i.e., how well it can handle previously unseen data. The network is therefore trained exclusively with the training dataset. At the end of the training, the test dataset is used to determine the test accuracy or accuracy of the neural network.

Unsupervised Learning

In contrast to supervised learning, no labeled data is required for unsupervised or self-supervised learning.

Unsupervised learning is about extracting patterns and structures from data without the need for labeled data. This type of learning typically includes tasks such as clustering, where similar data points are grouped, and dimensionality reduction, where high-dimensional data is represented in a low-dimensional space while preserving important information. Examples of unsupervised learning algorithms include K-Means clustering, hierarchical clustering, and Principal Component Analysis (PCA). Since they belong to the field of machine learning but do not represent artificial neural networks and thus are not the focus of this book, we do not want to delve further into the presentation of these methods (MacKay, 2003; Bishop & Nasrabadi, 2006).

Self-supervised Learning

Self-supervised Learning is on the one hand a special case of unsupervised learning and on the other hand a category of learning methods in its own right (Liu et al., 2021). In this case, the algorithm generates its own supervision signal (label) from the input data. For example, the model is trained to predict or reconstruct certain aspects of the data, e.g., the next image in a video or the context of a specific word in a sentence. In self-supervised learning, the data itself provides the labels or the supervision signal, with which the parameters of the model are updated during training. In this way, it is possible to learn from large amounts of unlabeled data, which are often much more extensive than labeled data. By learning to extract useful representations from the data, self-supervised learning can be used to improve the performance of a wide range of downstream tasks, including classification, object detection, and language understanding.

Often, the neural network is trained on an auxiliary task using the self-generated labels, which should help it, for example, in learning useful representations. After learning the auxiliary task, the learned representations can be used for downstream tasks.

An example of self-supervised learning is the task of predicting the next word in a sentence (or a masked word) in the context of natural language processing. In this case, the model generates its own supervision signal by using the surrounding words as context and thus learns useful language representations. In this way, so-called word vectors are generated, which form the basis of ChatGPT and Co. More on this in the chapter on language-talented AI.

Another example of self-supervised learning are autoencoders, which are also referred to as encoder-decoder networks. This type of neural networks consists of two parts. In the encoder, the layers from the input layer to the so-called bottleneck (Bottleneck Layer) become narrower with each further intermediate layer, thus containing fewer neurons. In the decoder, starting from the bottleneck, the layers successively become wider again up to the output layer, which has the same width (number of neurons) as the input layer (Fig. 12.5).

The idea behind this architecture is that the input is increasingly compressed by the encoder, while the decoder should be able to reconstruct the original input signal as accurately as possible from this compression. The compression must therefore be as lossless as possible. In the bottleneck layer, an abstract, reduced to the essentials representation of the input is created, which is also called embedding.

Autoencoders can be used to reduce noise in data or to complete incomplete data. The embeddings can also be read directly from the bottleneck layer and used for visualization or as input for further processing steps. A variant are autoencoders that are not trained end-to-end, i.e., completely over all layers, but instead layer by layer, with each layer having the task of compressing the input as well as possible. Once the training of a layer is completed, its weights are frozen, i.e., no longer changed, and its output serves as input for the next layer to be trained. This is an elegant way to solve the problem of vanishing gradients (Bengio et al., 1994), as only a shallow network is trained at a time. In this way, increasingly abstract feature and object representations are created with each additional layer.

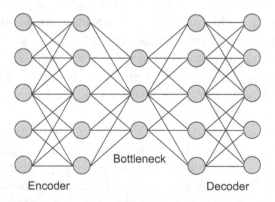

Fig. 12.5 Autoencoder. Schematic representation of an autoencoder, consisting of an encoder and a decoder part. The encoder compresses the input data into a narrower bottleneck, while the decoder reconstructs the data from this compressed representation. Compression and decompression should be as lossless as possible. The representation generated in the bottleneck, the so-called embedding, then represents a reduced to the essentials representation of the original input data

Neurons for Cats and Jennifer Aniston

In 2012, a Google team made an interesting discovery when it trained a deep autoencoder with a large dataset of images from YouTube videos. They found that one of the neurons in the network had learned to respond strongly to images of cats, even though the model had not been explicitly trained to recognize cats. Instead, it had learned this independently by finding patterns in the data (Le et al., 2012). The researchers subsequently named this neuron the "cat neuron".[1] In the human brain, too, there are neurons that react very selectively to certain concepts or objects. The so-called "Jennifer Aniston" neuron in the brain of an epilepsy patient, who had electrodes implanted in his brain before his operation to remove the epilepsy focus, gained fame in this context. This neuron reacted selectively to images of actress Jennifer Aniston, but not to images of other people (Quiroga et al., 2005).

Reinforcement Learning

In contrast to all previous learning methods, reinforcement learning is a type of machine learning where a model or agent is trained to learn useful input-output functions, i.e., to make a series of decisions in an uncertain environment that maximize a cumulative reward (Dayan & Niv, 2008; Li, 2017; Sutton & Barto, 2018; Botvinick et al., 2019; Silver et al., 2021).

In contrast to supervised learning, no outputs or labels are provided to the model. Instead, the agent receives feedback in the form of rewards or penalties after each action. The goal of the algorithm is to learn a strategy (Policy) that maps states of the environment to actions in such a way that it leads to a maximum reward in the long term.

The agent uses trial and error to learn from its experiences in the environment and randomly tries out different actions to find out which actions lead to the highest rewards in certain situations, by increasingly using actions that have already proven to be successful. Over time, the agent's strategy is continuously refined and optimized, so that it can make better decisions and achieve higher rewards.

---

[1] https://www.nytimes.com/2012/06/26/technology/in-A-big-network-of-computers-evidence-of-machine-learning.html

There are two types of reinforcement learning. In model-based reinforcement learning, a model of the environment is also learned, which can predict the feedback of the environment (rewards or penalties) to certain actions. In model-free reinforcement learning, on the other hand, the agent limits itself to learning the best action for a given state (Kurdi et al., 2019).

Reinforcement learning is used in a variety of applications, including games, robotics, autonomous driving, and recommendation systems.

The origins of reinforcement learning lie in psychology, particularly in the study of animal behavior and learning. The concept of reinforcement was first introduced in the 1930s and 1940s by B.F. Skinner, who developed the theory of operant conditioning (Skinner, 1963). Skinner's theory states that behavior is influenced by subsequent consequences such as reward or punishment.

Conclusion

Supervised learning requires labeled data to train models that can correctly classify input data such as images. In contrast, both unsupervised and self-supervised learning can process data without labels. In unsupervised learning, the focus is on discovering patterns and structures in the data, while self-supervised learning essentially generates its own labels from the data and thus learns useful representations by solving auxiliary tasks.

In the case of the autoencoder, each input pattern is simultaneously its own label, as it should be restored as accurately as possible from the compression of the bottleneck layer. The input and desired output are therefore identical. If the image of an apple is input, the desired output is this image of the apple. In predicting, for example, the next element to a certain input sequence (video or text), the next image or word is essentially the label of this sequence. Training is then again, as in the supervised case, with error backpropagation and gradient descent.

In reinforcement learning, training is focused on making decisions in an uncertain environment that maximize a cumulative reward, without using predefined outputs or labels. Instead, the agent uses trial and error to learn which actions yield the highest rewards, and refines its strategy over time. The concept originally comes from psychology and the study of animal behavior.

While we have made considerable progress in artificial intelligence and machine learning, it is important to emphasize that we are still at the beginning. Neuroscience has the potential to inspire new learning paradigms by giving us deep insights into the workings of the human brain. For example, insights from studies on brain plasticity could lead to new approaches for learning and adapting neural networks. Similarly, investigations into the mechanisms the brain uses to process complex sensory data could lead to new architectures for deep neural networks that are more efficient and robust. Furthermore, understanding the neural processes involved in decision-making and problem-solving could provide new approaches for reinforcement learning and optimization.

Finally, deeper investigations into the coordination and communication between different brain regions could contribute to the development of improved methods for training and coordinating larger ensembles of several different neural networks develop. Autoencoder with an encoder and a decoder part or Generative Adversarial Networks, which we will get to know more closely in the chapter on generative AI, are examples of ensemble networks made up of two subnetworks. In principle, however, ensembles of many subnetworks specialized in various tasks are also conceivable.

References

Bengio, Y., Simard, P., & Frasconi, P. (1994). Learning long-term dependencies with gradient descent is difficult. *IEEE transactions on neural networks, 5*(2), 157–166.

Bishop, C. M., & Nasrabadi, N. M. (2006). *Pattern recognition and machine learning* (Vol. 4, No. 4, p. 738). Springer.

Botvinick, M., Ritter, S., Wang, J. X., Kurth-Nelson, Z., Blundell, C., & Hassabis, D. (2019). Reinforcement learning, fast and slow. *Trends in Cognitive Sciences, 23*(5), 408–422.

Cybenko, G. (1989). Approximation by superpositions of a sigmoidal function. *Mathematics of Control, Signals and Systems, 2*(4), 303–314.

Dayan, P., & Niv, Y. (2008). Reinforcement learning: The good, the bad and the ugly. *Current Opinion in Neurobiology, 18*(2), 185–196.

Goodfellow, I., Bengio, Y., & Courville, A. (2016). *Deep learning.* MIT press.

Kurdi, B., Gershman, S. J., & Banaji, M. R. (2019). Model-free and model-based learning processes in the updating of explicit and implicit evaluations. *Proceedings of the National Academy of Sciences, 116*(13), 6035–6044.

Li, Y. (2017). *Deep reinforcement learning: An overview.* arXiv preprint arXiv:1701.07274.

Liu, X., Zhang, F., Hou, Z., Mian, L., Wang, Z., Zhang, J., & Tang, J. (2021). Self-supervised learning: Generative or contrastive. *IEEE Transactions on Knowledge and Data Engineering, 35*(1), 857–876.

Le, Q. V., Monga, R., Devin, M., Corrado, G., Chen, K., Ranzato, M. A., ... & Ng, A. Y. (2012). *Building high-level features using large scale unsupervised learning*. arXiv preprint. arXiv:1112.6209.

LeCun, Y., Bengio, Y., & Hinton, G. (2015). Deep learning. *Nature, 521*(7553), 436–444.

MacKay, D. J. (2003). *Information theory, inference and learning algorithms*. Cambridge University Press.

McCulloch, W. S., & Pitts, W. (1943). A logical calculus of the ideas immanent in nervous activity. *The Bulletin of Mathematical Biophysics, 5*, 115–133.

Quiroga, R. Q., Reddy, L., Kreiman, G., Koch, C., & Fried, I. (2005). Invariant visual representation by single neurons in the human brain. *Nature, 435*(7045), 1102–1107.

Rosenblatt, F. (1958). The perceptron: A probabilistic model for information storage and organization in the brain. *Psychological Review, 65*(6), 386.

Schmidhuber, J. (2015). Deep learning in neural networks: An overview. *Neural Networks, 61*, 85–117.

Silver, D., Singh, S., Precup, D., & Sutton, R. S. (2021). Reward is enough. *Artificial Intelligence, 299*, 103535.

Skinner, B. F. (1963). Operant behavior. *American Psychologist, 18*(8), 503.

Sutton, R. S., & Barto, A. G. (2018). *Reinforcement learning: An introduction*. MIT press.

13

Game-playing Artificial Intelligence

If there are intelligent beings on other planets, then they play Go.

Emanuel Lasker

Video Games

The development of reinforcement learning was a major breakthrough in artificial intelligence. In a groundbreaking work, the authors proposed combining reinforcement learning with so-called convolutional networks for image recognition, thereby learning directly from pixel values as input the actions to control a video game (Mnih et al., 2015). Specifically, the neural network received the last five images of the ongoing game and the score as input. The task was to generate the control commands for the controller or joystick, such as "left", "right", "up", "down" or "fire". The goal of the neural network was, as usual in reinforcement learning, to maximize the reward, i.e., the score.

The researchers trained their system on various Atari games and showed that a single algorithm without task-specific knowledge can achieve human-level performance. The AI system surpassed all previous methods of reinforcement learning and in some cases even the performance of human experts.

However, it turned out that the system was particularly good at action games that do not require a forward-looking strategy and where success does not depend on the previous course of the game, such as Video Pinball, Boxing or Breakout. All these games are pure reaction games, which can be won by purely tactical action (pressing the right button at the right time). In this type of game, the neural network reached human level or above. In contrast, the system completely failed at comparatively simple strategy games like Montezuma's Revenge or Ms. Pac-Man, where you have to navigate through a manageable world and solve certain tasks.

Nevertheless, the impact of the work on the development of AI was considerable. It marked a milestone in the research on reinforcement learning and showed that deep learning techniques can be combined with reinforcement learning algorithms to achieve unprecedented performance in a variety of tasks.

Go and Chess

The Asian board game Go, whose origins date back to ancient China, is considered the most complex strategy game ever. Its basic rules are relatively simple—much simpler than in chess—and can be learned within a few minutes. The complexity of the game arises from the sheer number of possible positions, which is estimated at 10^{170} and thus exceeds the number of possible positions in chess (approx. 10^{40}) by many orders of magnitude.

In recent decades, chess computers have achieved impressive performances and are now able to beat even the best human players. In contrast, Go computers until recently still had difficulties winning against even mediocre human players. This is because the brute-force method used in chess in Go is not applicable. In brute force, all possible moves for a certain number of moves in the future are simulated and the resulting positions are evaluated. The chess computer then selects the best move. Since there is a relatively limited number of moves and possible positions in chess, a computer with sufficient computing power can successfully apply this technique.

In contrast, brute force is ruled out in Go due to the astronomically high number of possibilities. There are more possible positions on the Go board than atoms in the universe, which means that even the most powerful computers currently would not be able to calculate all possible moves. Instead, Go computers must rely on a combination of heuristic techniques and machine learning to determine their moves.

Another difference between chess and Go is that chess can produce relatively simple positions after just a few moves, while Go can still be very complex even after many moves. After only two or four moves in chess,

Fig. 13.1   The game Go. The board game, whose origins date back to ancient China, is considered the most complex strategy game ever, although the basic rules are relatively simple, for example compared to chess. This is mainly due to the sheer number of possible positions, which exceeds that of chess by many orders of magnitude. Developing an AI that can master the game at an advanced human level was long considered unattainable

there are usually only a few possible positions, while in Go there are already thousands of possible positions after two or four moves. This makes Go an even greater challenge for computers, as they not only have to calculate the best moves, but also have to analyze and evaluate the complexity of the positions (Fig. 13.1).

AlphaGo Achieves the Breakthrough

While the supercomputer Deep Blue managed to defeat the then world chess champion Garry Kasparov for the first time in 1996, it was long considered unattainable to develop an AI that could master the game of Go at an advanced human level. However, this changed in 2016 with the AI system *AlphaGo* from the company DeepMind (Silver et al., 2016).

Initially, AlphaGo, a neural network, was trained on millions of historical Go games played by masters and grandmasters in tournaments, and on games against other computer programs. The network was supposed to predict the probabilities for different moves in a given situation. In addition, the network was used to analyze the game and evaluate possible moves and positions for both players. In a sensational match in March 2016, AlphaGo defeated the then best Go player in the world, Lee Sedol, in four out of five games, thus making history. For the first time ever, an AI had beaten a human at grandmaster level in Go. This spectacular success marked a milestone in the development of AI.

Just one year later, DeepMind presented *AlphaGo Zero* (Silver et al., 2017), a new approach to developing AI in the game of Go. Instead of training the network on historical games, DeepMind had AlphaGo Zero play billions of games against itself using reinforcement learning. At the beginning, the network was only given the rules, i.e., the allowed moves, and the goal of the network was to maximize the reward, i.e., the number of games won. At the beginning, AlphaGo Zero simply tried random moves, but it quickly improved.

After just four days of training, the new AlphaGo Zero was as good as AlphaGo and reached a level of play after another 40 days of training that was so superior that it could defeat its predecessor version AlphaGo—which had already beaten the human world champion—in 100 games just as often.

The worldwide Go elite set about analyzing AlphaGo Zero's style of play. Some grandmasters said that watching this AI play was like watching a hyper-intelligent alien play. AlphaGo Zero developed strategies, tactics, and moves that were completely unknown in the more than 4000-year history of the game and were significantly superior to the known ones in some cases. This AI has forever changed the game of Go. In the meantime, the analysis of human players' behavior against self-learning Go programs even resulted in a significant improvement in the abilities of the playing humans (Choi et al., 2022).

Now there is *AlphaZero* (Zhang & Yu, 2020), a further generalization of AlphaGo Zero. This system can teach itself any strategy game, such as chess, shogi (Japanese chess), checkers, and of course Go. The AlphaZero variant *AlphaStar* is even capable of playing the massive parallel online player strategy game *StarCraft II* at a human level against a variety of professional gamers (Jaderberg et al., 2019).

Conclusion

The success of AlphaGo and AlphaGo Zero demonstrates how far AI technology has come and what possibilities are opening up for the future. The complex game of Go serves as a vivid example of the impressive capabilities of artificial intelligence and the progress that has been made in this area.

However, cognitive scientist Gary Marcus points out that a large part of human knowledge has flowed into the development of AlphaZero. And he suggests that human intelligence seems to include some innate abilities, such as the intuitive ability to develop language. He advocates considering these innate abilities in the development of future AI systems, i.e., using a priori knowledge, instead of always starting training from scratch (Marcus, 2018).

Josh Tenenbaum, a professor at the Massachusetts Institute of Technology who also deals with human intelligence, argues similarly and says that we should study human flexibility and creativity if we want to develop real artificial intelligence at a human level. He highlighted, among other things, the intelligence of Demis Hassabis and his colleagues at the company DeepMind, who conceived, designed, and created AlphaGo in the first place (Lake et al., 2017).

References

Choi, S., Kim, N., Kim, J., & Kang, H. (2022). *How does AI improve human decision-making? Evidence from the AI-powered Go program. Evidence from the AI-powered Go program (April 2022).* USC Marshall School of Business Research Paper Sponsored by iORB, No. Forthcoming.

Jaderberg, M., Czarnecki, W. M., Dunning, I., Marris, L., Lever, G., Castaneda, A. G., … & Graepel, T. (2019). Human-level performance in 3D multiplayer games with population-based reinforcement learning. *Science, 364*(6443), 859–865.

Lake, B. M., Ullman, T. D., Tenenbaum, J. B., & Gershman, S. J. (2017). Building machines that learn and think like people. *Behavioral and Brain Sciences, 40*, e253.

Marcus, G. (2018). *Innateness, alphazero, and artificial intelligence.* arXiv preprint. arXiv:1801.05667.

Mnih, V., Kavukcuoglu, K., Silver, D., Rusu, A. A., Veness, J., Bellemare, M. G., … & Hassabis, D. (2015). Human-level control through deep reinforcement learning. *Nature, 518*(7540), 529–533.

Silver, D., Huang, A., Maddison, C. J., Guez, A., Sifre, L., Van Den Driessche, G., … & Hassabis, D. (2016). Mastering the game of Go with deep neural networks and tree search. *Nature, 529*(7587), 484–489.

Silver, D., Schrittwieser, J., Simonyan, K., Antonoglou, I., Huang, A., Guez, A., … & Hassabis, D. (2017). Mastering the game of Go without human knowledge. *Nature, 550*(7676), 354–359.

Zhang, H., & Yu, T. (2020). *AlphaZero. Deep reinforcement learning: Fundamentals, research and applications* (pp. 391–415). Springer.

14

Recurrent Neural Networks

Recurrent neural networks are universal parallel-sequential computers.

Jürgen Schmidhuber

Recurrence in the Brain

Recurrence is a fundamental aspect of neural processing and integration of information in biological neural networks, especially in the brain. As we have already seen in the chapter on the structure of the nervous system, numerous hierarchically nested feedback loops exist in the brain. Due to the total number of neurons in the brain (10^{11}) and the average number of connections per neuron (10^4) to its successors, an estimate suggests that each signal on average returns to its source after only three synapses (Braitenberg & Schüz, 1991): Since each neuron passes the signal to 10,000 other neurons, the signal after the first synapse is at 10^4, after the second synapse at $10^4 \times 10^4 = 10^8$ and after the third synapse theoretically at $10^4 \times 10^4 \times 10^4 = 10^{12}$ neurons, which however exceeds the actual number of existing neurons tenfold. Thus, the signal must now have returned to the original neuron among others. This highly recurrent structure allows the brain to process information extremely efficiently and dynamically.

The recurrent nature of biological neural networks plays an important role in many cognitive processes. For example, recurrent connections are crucial for the formation and retrieval of memories. When information

P. Krauss, *Artificial Intelligence and Brain Research*,
https://doi.org/10.1007/978-3-662-68980-6_14

flows through the network and returns to its source, it can strengthen the connections between the involved neurons, leading to the consolidation of memories. This process is facilitated by synaptic plasticity, i.e., the ability of synapses to change their strength over time depending on their activity.

In addition, the recurrent connectivity in the brain contributes to the integration of sensory information from different modalities, which is essential for perception and decision-making. For example, when we simultaneously see and hear something, the brain's recurrent circuits allow these separate inputs to be merged into a coherent representation of the external environment. This multisensory integration enables us to respond appropriately to our environment and make informed decisions.

In addition to memory formation and sensory integration, recurrent connections in the brain also play a crucial role for attention and consciousness. By selectively amplifying or inhibiting certain signals in feedback loops, the brain can modulate its attention to specific stimuli and allocate cognitive resources accordingly. This mechanism is important to filter out irrelevant information and maintain cognitive flexibility.

Recurrence in Artificial Neural Networks

In contrast to pure feedforward architectures, as we have learned in the previous chapter, artificial recurrent neural networks (RNN) also offer several advantages that make them suitable for a variety of tasks. One of the main advantages of RNNs is their ability to process temporal dependencies and sequences in the data, which is essential for tasks with time series data, natural language processing, and other problems where the order of the input is important.

In addition, RNNs can process or generate input and output sequences of variable length, while pure feedforward networks require input and output sequences of fixed size. Due to this flexibility, RNNs are better suited for tasks where the length of the input and output sequences varies, such as in language translation or speech recognition.

The recurrence can be expressed in different ways. In the simplest case, the network as a whole is a feedforward network, with the neurons of some or all layers having additional self-connections, so that in each time step of processing they receive their own output from the previous time step as input in addition to the input from the preceding layer. However, there are also network architectures where entire layers are fed back, up to fully recurrent neural networks, which consist almost entirely of a single layer of recurrently connected neurons.

LSTMs

Long-Short-Term Memories (LSTMs) are recurrent neural networks that are specifically designed to store information over long periods of time and selectively forget (Hochreiter & Schmidhuber, 1997). The core idea behind LSTMs is the use of a memory cell that can store information over a longer period of time. The memory cell is updated by a series of gate mechanisms that control the flow of information into and out of the cell. The gates are trained to learn which information to store, forget, or update at each time step. LSTMs have been successfully used for a variety of tasks, including speech recognition, machine translation, and automatic caption generation. They have also been used in combination with other neural network architectures such as convolutional networks to create more powerful models for tasks like object recognition and classification in images and video sequences.

Elman Networks

Elman Networks are another type of recurrent neural networks, proposed by Jeffrey Elman (Elman, 1990). In the simplest case, these are three-layer networks with an input, intermediate, and output layer, with the intermediate layer being extended by a so-called context layer. This context layer stores the state of the intermediate layer from the previous time step and passes it on to the intermediate layer. Thus, the intermediate layer receives the new input from the input layer and additionally its own activation state from the previous time step. Therefore, Elman Networks are also capable of processing input sequences and generating output sequences (Fig. 14.1).

Highly Recurrent Neural Networks

Hopfield networks are one of the most well-known examples and at the same time a special case of highly or fully recurrent neural networks (Hopfield, 1982). Although they are recurrent, Hopfield networks are not intended to process time series. However, they are briefly mentioned here as they exemplify the performance of neural networks with recurrent connections beyond the processing of sequences (Fig. 14.2).

Hopfield networks consist of a single layer of neurons, with each neuron i symmetrically connected to every other neuron j, i.e., for the weights

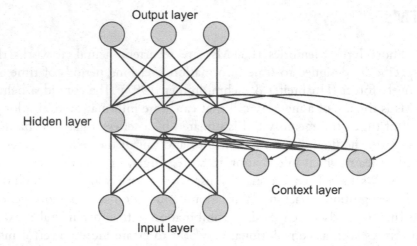

Fig. 14.1   Elman Networks. Named after their inventor Jeffrey Elman, they are a type of recurrent neural networks. They consist of three layers—input, intermediate, and output layer—and are characterized by an additional context layer that stores the state of the intermediate layer from the previous time step and passes it on. As a result, Elman Networks can process input sequences and generate output sequences

w, $w_{ij} = w_{ji}$ applies. Also, $w_{ii} = 0$, i.e., there are no self-connections. Hopfield networks exhibit pronounced attractor dynamics. Attractors are stable states towards which a dynamic system moves over time. Hopfield networks can store patterns as attractors and denoise or complete them upon re-presentation, i.e., the network activity converges (usually in one time step) into the attractor most similar to the input.

Highly recurrent neural networks, where the weights are no longer symmetric (reservoirs), represent complex dynamic systems, which are even capable of continuous activity without external input (Krauss et al., 2019).

As we will see in the next section, recurrent neural networks are difficult to train, with reservoirs being the most difficult to train. A radically new approach, motivated by findings in brain research (Reservoir Computing), even completely refrains from training these networks. But more on that later in Part IV.

Difficulty in Training Recurrent Networks

During learning, a phenomenon occurs in RNNs which is referred to as the problem of vanishing or exploding gradients (Hochreiter, 1998; Hanin, 2018; Rehmer & Kroll, 2020). This problem also occurs in a weakened form

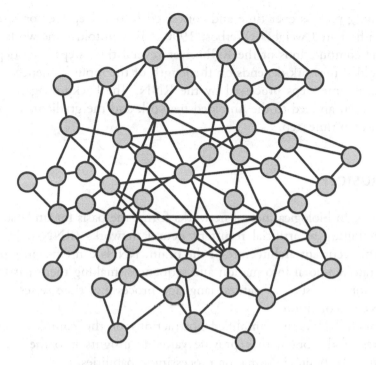

Fig. 14.2   Highly recurrent neural network. In this type of network, the ordered layer structure is completely abolished. In principle, each neuron can be connected to any other—even reciprocally. These networks are complex dynamic systems, which are capable of continuous activity even without input

in deep neural networks with pure feedforward architecture, but it is drastically exacerbated in RNNs. This refers to the instability of the gradients during error backpropagation, which can lead to slow or unstable learning. This problem occurs in two ways: vanishing and exploding gradients.

As the gradients are propagated backwards through the layers of the network, they can become very small with each additional layer (or with each additional time step in RNNs) and tend towards zero, leading to slow convergence, poor performance, and difficulties in learning long-range dependencies or meaningful features from the input data.

Exploding gradients occur when they grow exponentially during backpropagation, leading to unstable models, poor convergence or even divergence of the learning process, erratic behavior, and high prediction errors.

Unlike pure feedforward networks, RNNs can only be trained by a trick with gradient descent methods and error backpropagation. In backpropagation through time, the recurrent neural network is unfolded through the training process over time and converted into a deep feedforward network with shared weights (Werbos, 1990). This unfolded network represents the computations of the RNN over several time steps. The depth of the unfolded network depends on the length of the input sequence and the number of time steps processed by the RNN. The backpropagation algorithm is then applied to the unfolded network and the gradients are calculated for each time step.

Conclusion

Recurrence in biological neural networks forms the basis for understanding the functioning of artificial recurrent neural networks (RNNs). By mimicking the recurrent architecture of the brain, RNNs can efficiently process or generate temporal information and sequences, making them particularly suitable for tasks such as natural language processing, time series analysis, and speech recognition.

Although RNNs are simplified abstractions of the complex recurrent structures of the brain, they provide valuable insights into the principles underlying the brain's information processing capabilities.

However, recurrent networks are also more difficult to train than pure feedforward architectures. As we will see in the chapter on language-talented AI, recurrence has been completely replaced by another mechanism in large language models like ChatGPT, which is significantly easier and therefore faster to train.

References

Braitenberg, V., & Schutz, A. (1991). *Anatomy of the cortex: Studies of brain function*. Springer.

Elman, J. L. (1990). Finding structure in time. *Cognitive science, 14*(2), 179–211.

Hanin, B. (2018). Which neural net architectures give rise to exploding and vanishing gradients? *Advances in neural information processing systems, 31*.

Hochreiter, S. (1998). The vanishing gradient problem during learning recurrent neural nets and problem solutions. *International Journal of Uncertainty, Fuzziness and Knowledge-Based Systems, 6*(2), 107–116.

Hochreiter, S., & Schmidhuber, J. (1997). Long short-term memory. *Neural Computation, 9*(8), 1735–1780.

Hopfield, J. J. (1982). Neural networks and physical systems with emergent collective computational abilities. *Proceedings of the National Academy of Sciences, 79*(8), 2554–2558.

Krauss, P., Schuster, M., Dietrich, V., Schilling, A., Schulze, H., & Metzner, C. (2019). Weight statistics controls dynamics in recurrent neural networks. *PLoS ONE, 14*(4), e0214541.

Rehmer, A., & Kroll, A. (2020). On the vanishing and exploding gradient problem in gated recurrent units. *IFAC-PapersOnLine, 53*(2), 1243–1248.

Werbos, P. J. (1990). Backpropagation through time: What it does and how to do it. *Proceedings of the IEEE, 78*(10), 1550–1560.

15

Creativity: Generative Artificial Intelligence

Creativity means seeing what others see, and thinking what no one else has thought.

Albert Einstein

What is Creativity?

Before we turn to creative AI, let's first briefly define what we want to understand as creativity in the context of this presentation. Most definitions of creativity contain two main aspects. Creativity means creating something that is both new and useful. In addition, the results of Eagleman and Brandt suggest that creativity often arises from questioning old prejudices through three key methods: Bending, Blending, and Breaking. These concepts represent different strategies for manipulating existing ideas, concepts, or frameworks to create new insights, solutions, or creations (Eagleman & Brandt, 2017).

In Bending, an existing concept or idea is altered to create something new. By changing some aspects of the original concept while retaining its core, unique interpretations or innovations can be developed. Bending expands the boundaries of an idea or concept and opens up new perspectives and possibilities. In art, for example, Bending could mean taking a traditional painting style and adapting it to express a modern theme or object.

In Blending, two or more seemingly different ideas or concepts are combined to create a new, unified whole. By synthesizing elements from

P. Krauss, *Artificial Intelligence and Brain Research*,
https://doi.org/10.1007/978-3-662-68980-6_15

different sources, innovative connections and associations can emerge. The fusion can lead to the development of entirely new areas, such as in the case of bioinformatics, which combines biology and computer science. In literature, elements from different genres can be blended, for example, the combination of romance and science fiction into a new subgenre.

Breaking means dismantling or deconstructing an existing idea, concept, or framework to reveal the underlying components or assumptions. By questioning conventional knowledge or established norms, the way can be paved for new insights and breakthroughs. Breaking can lead to a reassessment of basic principles and promote the development of new theories, methods, or practices. In science, the process of breaking could mean questioning long-standing assumptions about a particular phenomenon, leading to a new understanding or explanation. Examples of this would be Einstein's theory of relativity or quantum mechanics.

The area of AI that deals with creativity, i.e., the generation of new content, is referred to as generative AI or generative deep learning and includes a variety of so-called generative models for generating images, videos, texts, spoken language, or music (Foster, 2019). The most important ones will be briefly introduced in the following.

Deep Dreaming: When the Input is Trained, Not the Network

Deep Dreaming is a method to generate new images with unique, dreamlike patterns and features (Mordvintsev et al., 2015). The process is based on the same neural networks used for image recognition tasks (convolutional networks), but works in reverse and optimizes the input image to generate certain patterns or features, rather than identifying them. So, it's not the network that is trained on a given input, but the input is adapted to a given network.

The process begins with a network that has already been pre-trained on a large number of images. Usually, a dataset containing millions of images and thousands of object classes has been used for this purpose. The network has therefore already learned to recognize various patterns and features from the training data through a hierarchical process, with early layers recognizing low-level features (edges, textures) and deeper layers recognizing high-level features (objects, scenes).

Next, the user selects a specific layer of the network they want to work with, thereby determining the level of abstraction of the newly generated image. Lower layers generate simpler patterns, while higher layers produce more complex features. The user provides an existing input image that will serve as the starting point for the Deep Dream process. This image is passed through all layers up to the selected layer. The activations in this layer are read out and represent the patterns and features recognized in the image at the corresponding level of abstraction.

Now, the user defines a target function, which usually aims to maximize the sum of the activations in the selected layer. This stimulates the neural network to enhance the recognized patterns and features in the input image. Similar to training with backpropagation, the gradient of the target function with respect to the input image is calculated. This indicates how the input image needs to be changed to increase the value of the target function and is applied to the input image to update it. This process is repeated for a certain number of iterations or until a termination criterion is reached. The resulting image contains enhanced patterns and features, giving it a unique, dream-like appearance.

As already mentioned, the degree of abstraction of the image content can be controlled by selecting the layer in the neural network, as early layers represent simpler patterns like corners and edges, while deep layers contain more abstract representations of whole objects or scenes. This leads to a paradoxically appearing curiosity. If a deep layer with abstract representations is chosen as the starting point of Deep Dreaming, the resulting images tend to be more figurative. In contrast, early layers with simpler representations lead to images that are reminiscent of abstract art.

Style Transfer

Style Transfer aims to apply the artistic style of one image (style image) to the content of another image (content image) to create a new image that combines the content of the first image with the style of the second image (Gatys et al., 2015, 2016). In the sense of the three strategies of creativity introduced above, this technique thus falls into the area of blending. For this purpose, the same networks are used again, which also serve for image recognition, to separate and recombine content and style information of images.

As with Deep Dreaming, the process begins with a network that has already been pre-trained on a large number of images. The user specifies the content image, which contains the motif or scene to be preserved, as well as the style image, which represents the artistic style to be applied to the content image. Subsequently, both images are passed separately through the neural network. For the content image, the activations from one or more late layers are used to capture high-level content features. For the style image, activations from multiple layers of all hierarchy levels are used to capture both low-level style features (e.g., textures) and high-level style features (e.g., structures). These style features are typically represented by so-called Gram matrices, which capture the correlations between different features in each layer and thus effectively encode the style information.

Subsequently, a starting image is first created, which often begins as random noise or as a copy of the content image. The goal is to iteratively update this starting image so that it corresponds to both the content features of the content image and the style features of the style image. This is achieved by defining a special error function, which consists of two main components: content error and style error. The content error measures the difference between the content features of the starting image and the content image, while the style error measures the difference between the style features of the starting image and the style image. The total error is the weighted sum of content and style error. By weighting the two parts, it can be determined how strong the respective influence on the new generated image should be, i.e., how far away it is from the original content or style. Again, the gradient of the total error is calculated, which indicates how the starting image needs to be changed to minimize the error. A corresponding optimization algorithm is applied iteratively to update the starting image. The final image retains the content of the content image, but now appears in the artistic style of the style image.

Generative Adversarial Networks

A Generative Adversarial Network (GAN) is a system of two coupled neural networks used to generate deceptively real images or videos, so-called Deep Fakes. It consists of a generator network and a discriminator network (Goodfellow et al., 2020).

The generator continuously creates new candidate images or videos, while the discriminator simultaneously tries to distinguish real images and videos from artificially generated ones. Over the course of training, both networks iteratively improve in their respective tasks. The Deep Fakes thus generated are usually indistinguishable from real images and videos.

Diffusion Models

Diffusion models are a class of generative models that produce images through a process known as Denoising Score Matching (Vincent et al., 2010; Swersky et al., 2011; Sohl-Dickstein et al., 2015). These models learn to generate images by simulating a diffusion process that transforms a target image into random noise, and then learn to reverse this process. The basic idea is to train the model to predict the statistical distribution of the pixel values of the original image from a noisy version.

The training of a diffusion model begins with a dataset of images. By adding Gaussian noise, a sequence of increasingly noisy versions of each image is then generated. This process is referred to as diffusion. Subsequently, the model is trained to predict or reconstruct the next less noisy version of the original image from a noisy version. During training, the model learns to predict the statistics of the original image at each step of the diffusion process. Through this training, the model becomes increasingly efficient at removing noise from images. When the trained model is now given random noise as input, it iteratively generates (hallucinates) a completely new random image from it.

Diffusion models typically consist of layers of neurons with local connectivity, with each neuron only connected to a small neighborhood of neurons in the previous layer. This local connectivity is advantageous for image processing as it allows the model to learn local patterns and capture spatial hierarchies.

To generate images from text descriptions, diffusion models can be combined with language models (see next chapter). There are various ways to achieve this coupling, but a common approach is the use of a technique known as conditional diffusion (Batzolis et al., 2021; Nichol et al., 2021). In this case, the diffusion model is made dependent on the text description by integrating it into the model architecture. First, the text description is encoded using a pre-trained language model (e.g., GPT-3 or BERT). This generates a high-dimensional vector representation of the meaning of the text (text embedding). The diffusion model is then conditioned on the text representation by modifying its architecture. This can be done by adding the text embedding as additional input or by integrating it into the hidden layers of the model. Finally, the model is trained with the same diffusion process as before, but now with images and the corresponding text descriptions, so the diffusion process does not run "freely", but is constrained by the respective text embedding.

If an image is to be generated from a text description, the conditioned model is again given random noise as visual input, but this time also with the encoding of the text from the language model. The model then generates a random image that corresponds to the input description.

The most well-known diffusion models include DALL-E 2, Stable Diffusion, and Midjourney.

DALL-E 2 (Dali Large Language Model Encoder 2) was specifically developed for the generation of high-quality photorealistic images from natural language descriptions. It uses a combination of image and text processing to generate abstract image representations from linguistic descriptions of objects, scenes, or concepts. These image representations are then used by a so-called decoder network to generate photorealistic images from them. DALLE-E 2 is freely accessible online.[1]

Released in 2022, *Stable Diffusion*[2] is also a generative model for generating detailed images from text descriptions. However, it can also be used for other tasks such as generating image-to-image translations based on a text prompt. What is special about Stable Diffusion is that its complete program code and all model parameters have been published.[3] It can be operated on most standard PCs or laptops equipped with an additional GPU and offers full access for systematic analysis, customization, or further development. This represents a departure from the practice of other AI models like ChatGPT or DALL-E 2, which are only accessible via cloud services and whose exact internal architecture has not been published so far.

Midjourney[4] is considered the most advanced diffusion model currently available. It is capable of even generating photorealistic images from text descriptions artificially.

Deep Fakes can also be generated with diffusion models. It has been shown that the newer diffusion models are significantly superior to the longer-existing Generative Adversarial Networks (Dhariwal & Nichol, 2021).

---

[1] https://openai.com/product/dall-E-2

[2] https://stablediffusionweb.com/

[3] https://github.com/CompVis/stable-diffusion

[4] https://www.midjourney.com.

Conclusion

There exists a whole range of various methods of generative AI for creating new content such as images, videos, or even music. The generated images are usually hardly distinguishable from real ones by humans. All such approaches fall into the areas of bending and blending, and probably not into the area of breaking.

It should not go unmentioned at this point that the creation of Deep Fakes is not limited to the creation of images. The system *VALL-E*[5] is capable of swapping the voice of any audio recording, for example. This is also a kind of style transfer, where the content of the spoken language is matched and preserved, while the target voice corresponds to the style that is being replaced (Wang et al., 2023).

By the way, the writing style of a text can also be swapped without changing the content. This is referred to as Prose Style Transfer. This falls into the domain of probably the most spectacular form of generative AI currently, that of the so-called large language models like ChatGPT, which due to their relevance and timeliness, a separate—the next—chapter is dedicated to.

References

Batzolis, G., Stanczuk, J., Schönlieb, C. B., & Etmann, C. (2021). Conditional image generation with score-based diffusion models. arXiv preprint arXiv:2111.13606.

Dhariwal, P., & Nichol, A. (2021). Diffusion models beat GANs on image synthesis. *Advances in Neural Information Processing Systems, 34*, 8780–8794.

Eagleman, D., & Brandt, A. (2017). *The runaway species: How human creativity remakes the world.* Catapult.

Foster, D. (2019). *Generative deep learning: Teaching machines to paint, write, compose, and play.* O'Reilly Media.

Gatys, L. A., Ecker, A. S., & Bethge, M. (2015). A neural algorithm of artistic style. arXiv preprint. arXiv:1508.06576.

Gatys, L. A., Ecker, A. S., & Bethge, M. (2016). Image style transfer using convolutional neural networks. In *Proceedings of the IEEE conference on computer vision and pattern recognition* (pp. 2414–2423).

---

[5] https://vall-e.io/

Goodfellow, I., Pouget-Abadie, J., Mirza, M., Xu, B., Warde-Farley, D., Ozair, S., … & Bengio, Y. (2020). Generative adversarial networks. *Communications of the ACM, 63*(11), 139–144.

Mordvintsev, A., Olah, C., & Tyka, M. (2015). *Inceptionism: Going deeper into neural networks*. Google Research Blog. https://research.google/pubs/pub45507.

Nichol, A., Dhariwal, P., Ramesh, A., Shyam, P., Mishkin, P., McGrew, B., … & Chen, M. (2021). *Glide: Towards photorealistic image generation and editing with text-guided diffusion models*. arXiv preprint arXiv:2112.10741.

Sohl-Dickstein, J., Weiss, E., Maheswaranathan, N., & Ganguli, S. (2015, June). *Deep unsupervised learning using nonequilibrium thermodynamics*. In International Conference on Machine Learning (pp. 2256–2265). PMLR.

Swersky, K., Ranzato, M. A., Buchman, D., Freitas, N. D., & Marlin, B. M. (2011). *On autoencoders and score matching for energy based models*. In Proceedings of the 28th international conference on machine learning (ICML-11) (S. 1201–1208).

Vincent, P., Larochelle, H., Lajoie, I., Bengio, Y., Manzagol, P. A., & Bottou, L. (2010). Stacked denoising autoencoders: Learning useful representations in a deep network with a local denoising criterion. *Journal of Machine Learning Research, 11*(12).

Wang, C., Chen, S., Wu, Y., Zhang, Z., Zhou, L., Liu, S., … & Wei, F. (2023). *Neural Codec Language Models are Zero-Shot Text to Speech Synthesizers*. arXiv preprint arXiv:2301.02111.

16

Talking AI: ChatGPT and Co.

I am an artificial intelligence that can process and generate answers in natural language to various types of questions and tasks.

ChatGPT

A Brief History of Natural Language Processing

The history of natural language processing (NLP) dates back to the mid-twentieth century, when early attempts at machine translation, such as the Georgetown-IBM experiment (see glossary), laid the foundation for this field (Booth & Richens, 1952).

In the following decades, NLP went through various phases, including rule-based systems that relied on manually created rules and linguistic knowledge, as well as statistical methods that used techniques such as Hidden Markov Models and Bayesian inference for language modeling (Rabiner, 1989). Significant advancements were made with the advent of machine learning. The introduction of word representation techniques further improved the capabilities of NLP (Mikolov et al., 2013). In recent years, the field has been revolutionized by deep learning techniques and the development of advanced architectures such as LSTMs (Hochreiter & Schmidhuber, 1997) and Transformers (Vaswani et al., 2017), ultimately leading to the development of large language models like BERT (Devlin et al., 2018) and GPT (Brown et al., 2020).

© The Author(s), under exclusive license to Springer-Verlag GmbH, DE, part of Springer Nature 2024
P. Krauss, *Artificial Intelligence and Brain Research*,
https://doi.org/10.1007/978-3-662-68980-6_16

Word Vectors

Techniques for representing words are crucial for natural language processing as they provide a way to convert text data into numerical representations that can be processed by machine learning algorithms. These representations capture the structure and semantics (meaning) of the language and allow models to recognize relationships between words and perform various tasks based on this.

One-Hot Encoding is a simple technique for representing words, where each word in the vocabulary is represented as a binary vector whose size (dimensionality) corresponds to the size of the vocabulary. The vector has a "1" at the index that corresponds to the position of the word in the vocabulary, and a "0" at all other positions. While one-hot encoding is easy to implement, it suffers from the curse of dimensionality, as the size of the vectors increases with the size of the vocabulary, leading to inefficient representations and increased computational complexity. Nevertheless, one-hot encoding is an important intermediate step in generating a dense encoding, i.e., word embeddings. These are low-dimensional continuous vector spaces in which the words are embedded (Latent Space Embeddings), as opposed to the high-dimensional and sparsely populated space of one-hot encoding.

Another advantage of word embeddings is that in this space, word vectors are arranged so that semantically similar words are closer together, while dissimilar words are further apart. Ideally, synonyms even occupy the same location. In the space of one-hot encoding, however, the location of a vector has no connection to the meaning of the corresponding word (Fig. 16.1).

Word2Vec is one of these neural network-based techniques for generating word embeddings (Mikolov et al., 2013, 2017). It learns word embeddings from large text corpora in an unsupervised manner, i.e., it does not require labeled data. Word2Vec is particularly effective at detecting semantic and syntactic relationships between words. There are two different approaches used in Word2Vec: the Continuous Bag of Words (CBOW) model and the SkipGram model.

In the CBOW architecture, the model learns word embeddings by predicting a target word based on its surrounding context words within a certain window. The input consists of the context words and the output is the target word. During training, the model adjusts the word embeddings to minimize the prediction error for the target word depending on the context words.

The SkipGram architecture is somewhat the reverse of the CBOW model. In this case, the model learns word embeddings by predicting the context words surrounding a given target word. The input is thus the target word, and the output is a context word within a certain window around the target word.

One-Hot Encoding

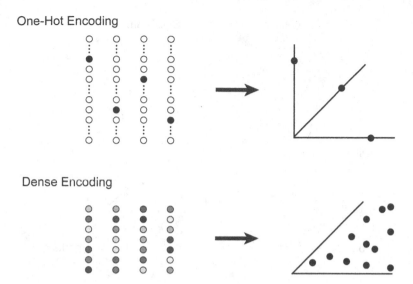

Dense Encoding

Fig. 16.1   One-Hot versus Dense Encoding. In one-hot encoding, each vector contains only one one, all other elements are zeros. The vector space has as many dimensions as there are words, with each word lying on its own coordinate axis. The vast majority of this incredibly high-dimensional space remains largely unoccupied. In dense encoding, on the other hand, significantly fewer dimensions are needed, resulting in a more compact vector space

Both the CBOW and the SkipGram model use a flat three-layer neural network with an input layer, a hidden layer, and an output layer. The word embeddings, which are adjusted during training to minimize prediction error, are created in the hidden layer. As the model learns, it captures the semantic and syntactic relationships between the words (Fig. 16.2).

The result is a more efficient and semantically rich representation of words, enabling NLP models to better understand and process language. By training Word2Vec on very large amounts of text, it learns the meaning of a word, so to speak, on the back of many other words that often appear with the word to be learned.[1] For example, in many texts, the words "bark",

---

[1] Learning the meaning of words on the back of other words is quite similar to a child's language acquisition. Although the first approximately 50 words are learned in a different way, namely by linking the respective sensory representation with the corresponding word. From the so-called 50-word limit, however, around the 2nd birthday, the so-called vocabulary spurt begins, with many new words being learned. It is assumed that this phase of language acquisition benefits from the fact that the meaning of new words can be inferred from already known words (Ferguson and Farwell, 1975; Rescorla, 1989; Aitchison, 2012).

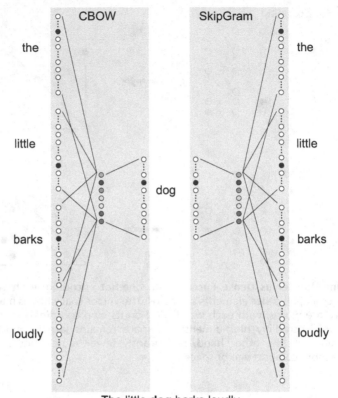

The little **dog** barks loudly.

Fig. 16.2 Word2Vec. Both the CBOW and the SkipGram model use a flat three-layer neural network with an input layer, a hidden layer, and an output layer. In CBOW, the context words serve as input and the target word as output. In SkipGram, it is exactly the opposite. Here, the target word is the input and the context words are the desired output. In both cases, the word embedding of the target word is created in the hidden layer, in this example "dog". In both cases, the dimensionality of the hidden layer is smaller than that of the input and output layers

"pet", and "fur" often appear together with "dog", thus defining its meaning. Conversely, the words "hamster", "cat", and "dog" often appear with the word "pet", thus defining its meaning, analogously for every other word in a language. The word embeddings generated by Word2Vec, e.g., v("dog"), then serve as input for further language processing algorithms such as machine translation or document classification.

A remarkable feature of word embeddings is the—initially surprising— fact that they can even be calculated with. For example, by adding and subtracting word vectors, new word vectors with meaningful meaning can be generated, as the following example impressively shows:

$$v(\text{"King"}) - v(\text{"Man"}) + v(\text{"Woman"}) = v(\text{"Queen"})$$

Transformer

Transformer (Vaswani et al., 2017) are the basis of all modern Large Language Models (LLM) like DeepL and ChatGPT. This new neural network architecture has revolutionized many tasks of natural language processing such as machine translation, question answering, and text summarization.

The Transformer architecture deviates from traditional recurrent neural networks (RNNs) by relying on a technique known as attention mechanism (Self-Attention). The attention mechanism allows input sequences to be processed in parallel rather than sequentially. This has the advantage that long-range dependencies—i.e., references between words with a greater distance within the sequence—can have a strong influence on the processing. If the sequence is processed serially as in RNNs, the influence of one word on another decreases very quickly with increasing distance. Another advantage of the attention mechanism is that it is highly parallelizable, which leads to shorter training times for the transformers (Fig. 16.3).

The clue behind the attention mechanism is that it allows the model to weight the meaning of different words in a specific context. To do this, transformers calculate a new representation for each word in the input sequence, which takes into account both the word itself and the surrounding context.

First, from the word embeddings (word vectors, x) of each word in the input sequence, three new vectors are calculated: query (q), key (k), and value vector (v). The values of the conversion matrices between the different types of vectors correspond to the internal parameters of the language model, which are learned during training.

From the query and key vectors, the attention (a) for each word at each position in the sequence is determined by calculating the respective scalar product. The attention indicates for each possible pair of words in the sequence how important one word is for understanding the other word in the sequence.

From the attention values of all words for a specific word and the corresponding value vectors of all words, a new representation (y) for this word is finally calculated, analogously for all other words. To do this, all value

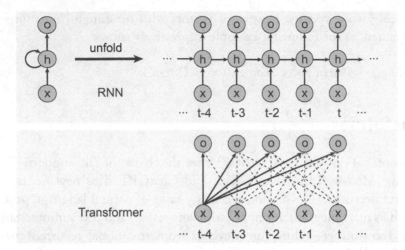

Fig. 16.3 RNN vs. Transformer. In an RNN, input sequences are processed serially. At each time step, the RNN receives the next word along with its own state in the previous time step as input. As a result, the influence of past words on the current word decreases very quickly with increasing distance in the sequence. In contrast, transformers process the sequence as a whole. This means that in principle, every word in the sequence can have a strong influence on every other. This is controlled by the attention mechanism

vectors are added, after they have been weighted by multiplication with the respective attentions (Fig. 16.4).

That one can calculate with the vectorial representation of words and thereby meaningful new meanings arise, we have already seen in the previous section using the example "King/Queen". The new representation of each word thus corresponds to its own meaning plus the weighted sum of the meaning of all other words in the specific context of the entered text. Simplified, the following happens: The "standard meaning" of each word, as it would be in a dictionary, is modified and adapted to the specific situation. From a linguistic point of view, this corresponds to the transition from semantics to pragmatics.

Modern transformers are made up of many such modules, consisting of attention and subsequent forward-directed neural network. Like the layers in a deep neural network, they are connected in series, with the output of one module serving as input for the subsequent module serves. ChatGPT, for example, consists of 96 of these modules or layers. In addition, on each layer there are several of these modules in parallel, which can then specialize in different aspects, such as different languages or types of text.

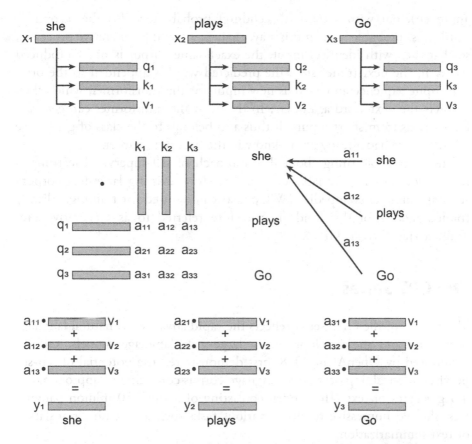

Fig. 16.4 Attention mechanism. From each word vector x of the input sequence, three new vectors are generated: query (q), key (k), and value vector (v). The attention a, which represents the relative importance of a word for understanding another word in the sequence, is calculated by the scalar product of query and key vectors. Finally, a new representation for each word y is created by weighting the value vectors of all words with the respective attention and adding them up

It has been shown that by iterative prediction of the next word based on the previous input sequence of words and subsequent appending of the predicted word to the next input, in principle arbitrarily long, meaningful texts can be generated (Liu et al., 2018). The final output, which the transformer is trained on, therefore corresponds to the probability for each individual word of the language (which can be several tens of thousands) that it will appear next in the text. Subsequently, a word is randomly selected according to this probability distribution over all words. So it is not generally the most probable word that is chosen (which would correspond to a winner-takes-all approach), but with certain (descending) probabilities also the second or third most probable, etc. In this way, chance enters the generative processes, so that even with identical input, the exact same output is never produced twice. In the next (time) step, the predicted word is appended to the original input sequence and used as new input for the transformer, which then predicts the next word again, etc. In this way, the transformer can generate longer texts from short inputs. It thus also belongs to the class of generative AI, which we had already got to know in the previous chapter.

The remarkable thing about this approach of self-supervised learning is that it does not require labeled data. Therefore, existing large text corpora such as online encyclopedias (Wikipedia) can be used for training. Already trained models of this kind are therefore referred to as Generative Pre-Trained Transformer (GPT).

The GPT Series

The history of the GPT series reflects the rapid advances in natural language processing and the development of large-scale language models. GPT-1, introduced by OpenAI in 2018, first demonstrated the potential of self-supervised pre-training on a large language corpus consisting of approximately 4.5 gigabytes of text. The model, consisting of about 150 million parameters, showed impressive results in various tasks such as machine translation or text summarization.

GPT-2, released in 2019, built on the success of its predecessor and significantly expanded the scope and capabilities of the model. With 1.5 billion internal parameters, GPT-2 was trained on a larger dataset (40 gigabytes) and was able to generate even more coherent and contextually relevant texts. Due to concerns about potential misuse, OpenAI initially withheld the release of the full model and decided to only release smaller versions, to publish the full model at a later date after a risk assessment.

GPT-3, introduced in 2020 and trained on 570 gigabytes of text, was another major step in the development of large language models. With 175 billion parameters, GPT-3 demonstrated remarkable learning ability in a few steps, with the model performing very well in various tasks with minimal fine-tuning. It is capable of inventing stories in various styles, can write short computer programs in all common programming languages, summarize documents, and translate texts. GPT-3 has already been integrated into numerous applications, including chatbots and so-called co-pilot tools for code completion. ChatGPT is also based on this version of the GPT series.

Finally, in March 2023, GPT-4 was released, which in some tests apparently performs even better than humans and shows signs of general artificial intelligence (Bubeck et al., 2023). So far, no details have been published about its exact structure and the training data used.

ChatGPT

The freely accessible ChatGPT, released in November 2022, is probably the most well-known example of a large language model and one of the most advanced AI models for conversations. It is based on GPT-3 and is designed to understand and generate human-like responses in a dialogue environment. It is so powerful that it can conduct longer coherent and context-related conversations with users. It can generate any type of text in seconds, answer questions on any topic, and conduct conversations, remembering their course and thus usually responding adequately in longer dialogues. ChatGPT can summarize, rephrase, or translate texts in several dozen languages, tell jokes, write songs, and even program in all common programming languages.

Remarkably, ChatGPT is even capable of playing strategic games like chess, Go, and poker. During its training on virtually any type of text available on the internet, it has, for example, "read" tens of thousands of games in chess notation, each time trying to predict the next word or character. This apparently resulted in the side effect that it learned the underlying rules of the respective game without ever having seen a chessboard or pieces. This suggests that the approach of encoding a problem into a sequence of elements and then learning to predict the next element could be a universally applicable strategy in information processing and cognitive systems.

As already mentioned, ChatGPT is based on GPT-3. Unlike its predecessors, which were trained exclusively self-supervised (on predicting the next word), ChatGPT underwent subsequent fine-tuning through human feedback, which proved to be a crucial step for generating even better responses and texts.

This fine-tuning was based on supervised learning and reinforcement learning (Reinforcement Learning) and proceeded in three steps. First, humans generated many example conversations, each consisting of a request and the corresponding appropriate responses. To assist them in designing their responses, the human trainers could access previously automatically generated suggestions and then adapt them. In this way, a large dataset was created, which contains real dialogues with good, i.e., adequate responses.

To further enlarge the dialogue dataset, artificially generated examples from a predecessor of ChatGPT were added. These example dialogues were used to train ChatGPT in a supervised manner.

For the subsequent reinforcement learning, rewards are required, which evaluate the quality of the output generated by the AI system. To provide these rewards for a large number of training examples automatically, a separate reward model was trained—another neural network, which receives a request along with a possible response as input and whose task is to predict the quality of the response as output, which is then used as a reward for fine-tuning ChatGPT.

To train the reward model, training data had to be created, of course. These data consisted of the responses generated by the chatbot, which were sorted by their quality. For this purpose, conversations between human AI trainers and the chatbot were collected, and several alternative additions were created for each model response. The human AI trainers then had the task of evaluating the different responses. With these evaluation data, the reward model could be trained to predict an appropriate evaluation (reward) for new response options that it had not seen before.

The fine-tuning of ChatGPT proceeded in such a way that a random request from the dialogue dataset was given to ChatGPT as input. ChatGPT then generated a response. The response was evaluated by the reward model, and this evaluation was reported back to ChatGPT as a reward to train it with reinforcement learning, i.e., to maximize the total reward. This improved its ability to generate high-quality context-related responses in a conversation situation. Through this iterative improvement process, ChatGPT became an increasingly efficient conversation AI over time (Fig. 16.5).

Speech Interfaces

Speech interfaces such as Siri, Alexa, and Google Assistant have revolutionized the way we interact with our devices. They provide a conversational interface that responds to user requests in a natural, human-like manner. These voice-controlled AI assistants are trained with large amounts of text data, like ChatGPT and other language models, to understand and generate natural language. While Siri, Alexa, and Google Assistant primarily focus on responding to specific commands and questions, often with a focus on functional support, e.g., setting reminders or playing music, models like ChatGPT are designed to generate more nuanced and context-related

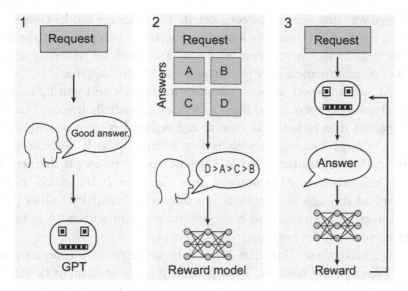

Fig. 16.5 Fine-tuning through human feedback. Step 1: A request is randomly selected from the database. An AI trainer provides an example response, with which GPT is trained. Step 2: A request and several example responses are randomly selected from the database. An AI trainer evaluates the quality of the responses. This trains the reward model. Step 3: A new request is selected from the database and given to GPT as input, whereupon it generates a response. The reward model determines the quality of the response. This is reported back to GPT as a reward for reinforcement learning

responses, enabling more complex and open conversations. Nevertheless, the underlying technology is similar.

Ultimately, the translation from written to spoken language and vice versa (Text-to-Speech, Speech-to-Text) represents just another recoding step, which already works very efficiently, as the example of Siri and Co. shows (Trivedi et al., 2018). Coupling this type of language interfaces with ChatGPT or GPT-4 is the next logical (and not particularly large) step and would make communication with these AI systems even more intuitive and natural.

Conclusion

The Transformer architecture and large language models like the GPT series and DeepL have revolutionized natural language processing, with the attention mechanism and the ability to consider far-reaching dependencies playing a key role. Despite these advances, there are still challenges and limitations that need to be considered. The language model *Galactica* of the Meta corporation, formerly known as Facebook, was taken off the net at the end of 2022 due to criticism and concerns about its reliability and the spread of misinformation after just three days.[2] What happened?

The AI was supposed to assist scientists in research and writing technical articles. However, it was found that Galactica had partially invented content, but presented it as factual and even mixed real and false information. This behavior of large language models is also referred to as hallucinating. The massive criticisms eventually led to Galactica being taken off the net. This incident reminds us of Microsoft's chatbot *Tay* from 2016, which continuously evolved through user requests and due to its sensitivity to user preferences, turned into a racist and homophobic program within 16 hours and had to be taken off the net.[3]

Examples like these show that AI models can sometimes generate unreliable or even harmful content. These models can invent content or mix true and false information, leading to misinformation and unwanted behavior. ChatGPT is not free from errors and can generate plausible-sounding, but false or nonsensical responses. It also reacts sensitively to changes in the input formulation and tends to be verbose or use certain expressions too frequently. It also tends to guess in the case of ambiguous requests, rather than asking clarifying questions, and although efforts have been made to get the model to reject inappropriate requests, it can still respond to harmful instructions or show biased behavior. Its successor GPT-4 is already significantly more reliable and advanced in this respect, but certainly still not free from such errors.

References

Aitchison, J. (2012). *Words in the mind: An introduction to the mental lexicon.* Wiley.
Booth, A. D., & Richens, R. H. (1952). *Some methods of mechanized translation.* In Proceedings of the Conference on Mechanical Translation.
Brown, T., Mann, B., Ryder, N., Subbiah, M., Kaplan, J. D., Dhariwal, P., & Amodei, D. (2020). Language models are few-shot learners. *Advances in Neural Information Processing Systems, 33,* 1877–1901.

---

[2] https://www.technologyreview.com/2022/11/18/1063487/meta-large-language-model-ai-only-survived-three-days-gpt-3-science/

[3] https://www.nytimes.com/2016/03/25/technology/microsoft-created-A-twitter-bot-to-learn-from-users-it-quickly-became-A-racist-jerk.html

Bubeck, S., Chandrasekaran, V., Eldan, R., Gehrke, J., Horvitz, E., Kamar, E., … & Zhang, Y. (2023). *Sparks of artificial general intelligence: Early experiments with GPT-4*. arXiv preprint arXiv:2303.12712.

Devlin, J., Chang, M. W., Lee, K., & Toutanova, K. (2018). *Bert: Pre-training of deep bidirectional transformers for language understanding*. arXiv preprint arXiv:1810.04805.

Ferguson, C. A., & Farwell, C. B. (1975). Words and sounds in early language acquisition. *Language*, 419–439.

Hochreiter, S., & Schmidhuber, J. (1997). Long short-term memory. *Neural Computation, 9*(8), 1735–1780.

Liu, P. J., Saleh, M., Pot, E., Goodrich, B., Sepassi, R., Kaiser, L., & Shazeer, N. (2018). *Generating wikipedia by summarizing long sequences*. arXiv preprint arXiv:1801.10198.

Mikolov, T., Chen, K., Corrado, G., & Dean, J. (2013). *Efficient estimation of word representations in vector space*. arXiv preprint arXiv:1301.3781.

Mikolov, T., Grave, E., Bojanowski, P., Puhrsch, C., & Joulin, A. (2017). *Advances in pre-training distributed word representations*. arXiv preprint arXiv:1712.09405.

Rabiner, L. R. (1989). A tutorial on hidden Markov models and selected applications in speech recognition. *Proceedings of the IEEE, 77*(2), 257–286.

Rescorla, L. (1989). The Language Development Survey: A screening tool for delayed language in toddlers. *Journal of Speech and Hearing disorders, 54*(4), 587–599.

Trivedi, A., Pant, N., Shah, P., Sonik, S., & Agrawal, S. (2018). Speech to text and text to speech recognition systems? A review. *IOSR Journal of Computer Engineering, 20*(2), 36–43.

Vaswani, A., Shazeer, N., Parmar, N., Uszkoreit, J., Jones, L., Gomez, A. N., … & Polosukhin, I. (2017). Attention is all you need. Advances in neural information processing systems, *30*.

17

What are AI Developers Researching Today?

It is no shame to know nothing, but rather, not wanting to learn anything.

Plato

Learning to Learn

In the rapidly evolving world of Artificial Intelligence, Machine Learning in particular has made great strides in recent years. Almost daily, technical articles are published introducing new methods or further developing existing ones. Even experts often find it difficult to keep up with the enormous speed of development.

In this chapter, we want to take a closer look at some of these latest trends and developments. All these innovative approaches have in common that they have the potential to revolutionize the way AI systems learn and adapt. The conventional approach of supervised learning requires a large amount of labeled data for each new task, the procurement of which can be time-consuming and expensive. The new approaches are intended to enable efficient use of prior knowledge, generalize across tasks, or solve novel problems with minimal training.

Few-Shot, One-Shot, and Zero-Shot Learning are some of these advanced Machine Learning techniques that are supposed to enable models to learn new tasks or recognize objects with minimal amounts of data. These approaches

P. Krauss, *Artificial Intelligence and Brain Research*,
https://doi.org/10.1007/978-3-662-68980-6_17

have gained importance in recent years as they can potentially overcome the need for huge amounts of training data and shorten training time.

Few-Shot Learning

Humans are very good few-shot learners: You don't have to show a child thousands of pictures of apples for it to learn the concept of "apple". A few, often even a single example, is usually sufficient.

With Few-Shot Learning, the aim is to train models that can quickly adapt to new tasks with a small amount of training data, such as classifying new objects based on a few examples (Snell, 2017; Sung et al., 2018). To do this, the model is first trained on a relatively small dataset that contains only a few examples for each class or task, and then tested on a new set of examples. The idea is to teach the model to learn from a few examples and generalize to new examples, rather than needing large amounts of data for each task. A recent development in this area are Prototypical Networks (Snell et al., 2017), where a deep neural network is used to learn a metric space in which objects of the same class are grouped. This method has shown promising results in tasks such as image classification, where the model can recognize new categories with just a few examples.

One-Shot Learning

With One-Shot Learning, the concept of learning from a few examples is continued by having the model learn from just one example per class (Santoro et al., 2016; Vinyals, 2016). A notable innovation in this area are Memory Enhanced Neural Networks. This type of neural network uses an external memory matrix to store and retrieve information about previously seen examples, so that the model can make accurate predictions based on a single example. One-Shot Learning has proven particularly useful in tasks such as handwriting recognition, where a model can accurately recognize a writer's style based on a single example.

In biology, One-Shot Learning can be observed, for example, in animals that are able to quickly recognize and respond to new stimuli or situations without having been previously exposed to them. For instance, some bird species are able to quickly recognize and avoid dangerous prey after a single experience. Of course, humans are also excellent One-Shot learners.

Zero-Shot Learning

With Zero-Shot Learning, on the other hand, models can make predictions for completely unknown classes without explicit training examples (Norouzi et al., 2013; Socher et al., 2013). The model is trained to recognize objects or categories it has never seen before. It can therefore classify new input patterns even if no labeled data for the respective class were available during training. In contrast to supervised learning, where a model is trained with a certain amount of labeled data examples, Zero-Shot Learning is based on the transfer of knowledge from related or similar classes that were seen during training.

This is achieved by using semantic representations such as word vectors that capture the meaning and relationships between different classes. In this case, the word vectors replace the labels of the images. For example, if a model has been trained to recognize images of different animal species (e.g., horse, tiger, ...), but has never seen an image of a zebra, it can still classify it as an animal and categorize it into a new category, as it has learned the relationships between different animal species. In this case, the zebra would probably be classified into a mixed category of the already learned categories "horse" (because of the shape of the animal) and "tiger" (because of the stripes).

Zero-Shot Learning thus enables more efficient and flexible training of models and generalization to new and unknown categories.

Transfer Learning

In transfer learning, a neural network is first trained on a very large dataset and then refined on a smaller dataset for a specific task or special application. This retraining of an already trained network is called fine-tuning. The idea behind it is that the knowledge acquired in solving one problem can be transferred to another, related problem, thereby reducing the amount of data and the time required for training a new model (Torrey & Shavlik, 2010).

If the neural network was initially trained on the gigantic ImageNet dataset, which consists of 14 million images divided into 20,000 categories, it can be assumed that this network has already learned a lot of representations that are useful for general image recognition. Thus, this network can be efficiently adapted to a new task with a short fine-tuning on a few images.

Meta-Learning

In traditional machine learning, a model is trained on a fixed set of training data and then used to make predictions for new, unseen data. In contrast, meta-learning approaches aim to enable models to learn from a small amount of data and generalize to new tasks with little or no additional training (Finn et al., 2007; Santoro et al., 2016). To this end, a so-called meta-learner is usually trained, i.e., a model that learns how to learn by observing and extracting patterns from a series of different training tasks. The meta-learner then uses this knowledge to quickly adapt to new tasks with similar features.

There are various approaches to meta-learning, including metric-based learning, optimization-based learning, and model-based learning. In metric-based learning, a similarity metric is learned to compare new tasks with the training tasks. In optimization-based learning, a model is trained to quickly optimize its weights for a new task, while in model-based learning, a generative model of the data is learned that can be used for quick adaptation to new tasks. For example, a meta-learner can be generally trained to play various strategy games or card games. In a new game, it can then use its existing knowledge about how games generally work to learn the new game faster. A similar effect is also observed in translation software.

In 2016, Google introduced the Google Neural Machine Translation (GNMT) system, a significant advancement in the field of machine translation that surpassed all existing models (Wu et al., 2016). An interesting observation was made when the GNMT system, which was initially trained to translate from English to Spanish and then to translate from English to Chinese, improved its translation between English and Spanish after being trained with Chinese text!

This phenomenon can be attributed to the development of a so-called interlingua, i.e., a common representation of meaning between languages. As the model was trained on more language pairs, it learned to map the input text from various languages onto a semantic space that was independent of the specific language. As a result, the model could now also translate between languages, even if it had not been explicitly trained for this specific language pair. The latest version of GNMT now supports 109 languages and can translate between them in any direction, which is over 11,000 possible language combinations. The model was explicitly trained only on a very small fraction of all these possibilities.

The emergence of this interlingual representation in the GNMT system is an impressive example of meta-learning and demonstrates the ability of deep learning models to learn abstract features that can be generalized across different tasks. This observation inspired further research into multilingual and zero-shot translation models, which aim to leverage shared knowledge between languages and improve translation performance with fewer training examples for each language pair.

Incidentally, it is also known from human language learners that the "cognitive effort" for each additional foreign language becomes somewhat smaller. We never start learning from scratch.

Hybrid Machine Learning

In hybrid machine learning, various techniques, e.g., of deep learning, are combined with other concepts (Maier et al., 2022). An example is Known Operator Learning, i.e., learning with known operators. In this case, individual layers of a neural network are replaced by so-called operators, i.e., mathematical functions, e.g., a Fourier transformation. This inclusion of prior knowledge about how the data needs to be transformed also has the advantage of accelerating learning and requiring fewer large datasets (Maier et al., 2019).

In the research direction of neurosymbolic AI, the strengths of neural networks and symbolic reasoning from mathematics and logic are combined (De Raedt, et al., 2020; Sarker et al., 2021). While the neural networks are used as usual for pattern recognition and learning regularities and patterns from large amounts of data, symbolic reasoning is used for logic, knowledge representation, and decision-making. According to Gary Marcus, this is one of the keys to human-like AI (Marcus, 2003).

Conclusion

The mentioned methods open up new possibilities for AI systems to learn more efficiently and adapt to new situations, thus paving the way for more versatile and robust AI applications in a wide range of application areas. Beyond the presented methods, such as how machines learn to learn, intensive research is already being conducted on the next steps of development.

Deep neural networks can be compared to highly specialized computer programs. However, understanding and interpreting these programs can pose a significant challenge. In order to better identify and decipher the patterns and operations hidden in deep networks, researchers have begun to transfer methods from software programming (Maier et al., 2022).

References

De Raedt, L., Dumancic, S., Manhaeve, R., & Marra, G. (2020). *From statistical relational to neuro-symbolic artificial intelligence.* arXiv preprint arXiv:2003.08316.

Finn, C., Abbeel, P., & Levine, S. (2017). *Model-agnostic meta-learning for fast adaptation of deep networks.* In International Conference on Machine Learning (pp. 1126–1135). PMLR.

Maier, A., Köstler, H., Heisig, M., Krauss, P., & Yang, S. H. (2022). *Known operator learning and hybrid machine learning in medical imaging – a review of the past, the present, and the future.* Progress in Biomedical Engineering.

Maier, A. K., Syben, C., Stimpel, B., Würfl, T., Hoffmann, M., Schebesch, F., & Christiansen, S. (2019). Learning with known operators reduces maximum error bounds. *Nature Machine Intelligence, 1*(8), 373–380.

Marcus, G. F. (2003). *The algebraic mind: Integrating connectionism and cognitive science.* MIT press.

Norouzi, M., Mikolov, T., Bengio, S., Singer, Y., Shlens, J., Frome, A., ... & Dean, J. (2013). *Zero-shot learning by convex combination of semantic embeddings.* arXiv preprint arXiv:1312.5650.

Santoro, A., Bartunov, S., Botvinick, M., Wierstra, D., & Lillicrap, T. (2016, June). *Meta-learning with memory-augmented neural networks.* In International conference on machine learning (pp. 1842–1850). PMLR.

Sarker, M. K., Zhou, L., Eberhart, A., & Hitzler, P. (2021). Neuro-symbolic artificial intelligence. *AI Communications, 34*(3), 197–209.

Socher, R., Ganjoo, M., Manning, C. D., & Ng, A. (2013). *Zero-shot learning through cross-modal transfer.* Advances in neural information processing systems, 26.

Snell, J., Swersky, K., & Zemel, R. (2017). *Prototypical networks for few-shot learning.* Advances in neural information processing systems, 30.

Sung, F., Yang, Y., Zhang, L., Xiang, T., Torr, P. H., & Hospedales, T. M. (2018). *Learning to compare: Relation network for few-shot learning.* In Proceedings of the IEEE conference on computer vision and pattern recognition (pp. 1199–1208).

Torrey, L., & Shavlik, J. (2010). Transfer learning. In*Handbook of research on machine learning applications and trends: algorithms, methods, and techniques* (pp. 242–264). IGI global.

Vinyals, O., Blundell, C., Lillicrap, T., & Wierstra, D. (2016). *Matching networks for one shot learning.* Advances in neural information processing systems, 29.

Wu, Y., Schuster, M., Chen, Z., Le, Q. V., Norouzi, M., Macherey, W., ... & Dean, J. (2016). *Google's neural machine translation system: Bridging the gap between human and machine translation.* arXiv preprint arXiv:1609.08144.

Part III
Challenges

Where there is much light, there is naturally often just as much shadow. Amidst all the spectacular successes in the field of Artificial Intelligence, particularly in Deep Learning, it should not be overlooked that there are many smaller and some fundamental challenges—some even speak of major crises—which still need to be solved.

Of course, brain research is still far from having developed an overarching theory of the brain, as we have seen particularly in the chapters on consciousness and free will. But even apart from these very big questions, there are still many aspects of how the brain works that remain ununderstood.

In this third part of the book, the aim is to describe the major challenges of both disciplines.

18

Challenges of AI

Realistically speaking, deep learning is just a part of the larger challenge associated with building intelligent machines.

Gary Marcus

It's All About the Data

The US Army had a plan to use neural networks to automatically identify hidden and camouflaged tanks. The Pentagon commissioned an external software company for the project. The company designed a neural network that was trained with 50 images of camouflaged tanks located among trees, as well as 50 images of wooded areas without tanks. After the training, the network was tested with 50 additional images per category, which it had not seen before. Indeed, all images of the test dataset were correctly classified and the neural network was submitted to the Pentagon.

However, it soon claimed that the neural network was no better than chance at distinguishing between images with and without tanks. Upon closer examination, it turned out that the dataset used for training and testing the network contained only photos of camouflaged tanks on cloudy days and of forests without tanks on sunny days. Therefore, the neural network did not learn to distinguish between camouflaged tanks and empty forests, but instead to differentiate between sunny and cloudy days (Yudkowsky, 2008).

© The Author(s), under exclusive license to Springer-Verlag GmbH, DE, part of Springer Nature 2024
P. Krauss, *Artificial Intelligence and Brain Research*,
https://doi.org/10.1007/978-3-662-68980-6_18

This example impressively shows that artificial intelligence can only be as good as the data it was trained with. While the inadequacy of the AI system in this case was noticed in time, such a bias in the training data, i.e., a systematic distortion or unintended correlation between features, can have dramatic consequences if it is not detected in time.

Autonomous driving systems like the Tesla Autopilot also rely on deep neural networks for image processing, for example to recognize traffic signs, other vehicles, obstacles or the road boundary and to derive the appropriate action from it. In the early days, a Tesla driver died during autopilot operation after the vehicle raced at full speed into a truck trailer instead of braking or evading.[1] An analysis of the software revealed that Tesla's image recognition interpreted the semi-trailer crossing the intersection as a bridge. The software had never seen a truck trailer from the side during its training. And the best fitting category, i.e., the most similar pattern, was a bridge. Since one can drive under bridges without danger, the appropriate action from the autopilot's point of view was to continue at the same speed.

So the data should not only contain as little bias as possible, but above all they should be sufficiently complete and not contain any critical white spots.

Hallucinating Chatbots

In this context, it is also worth mentioning again the problem of hallucinating chatbots *(ChatGPT, Galactica)* and their potential susceptibility to producing extremist content *(Tay)*, both of which we have already encountered in the chapter on linguistically gifted AI. Obviously, these models lack some kind of world model or internal fact checker. This also represents an unsolved challenge.

Dangerous Stickers and Other Attacks

While the previous examples show how important the quality of training data is for the performance of the AI system, the next examples show how AI systems can be deliberately misled.

In Fig. 18.1, what do you think is the difference between the left and the middle picture?—You don't see any?

---

[1] https://www.theverge.com/2016/6/30/12072408/tesla-autopilot-car-crash-death-autonomous-model-s

Fig. 18.1   Adversarial Example. For a human, there is no noticeable difference between the left and the middle picture. However, a network trained on image recognition will correctly classify the left picture as a "panda", while it assigns the middle picture to the category "gibbon". The middle picture is an adversarial example, which was artificially created to deceive the classifier. To do this, a specially created pattern (right) was added to the pixels of the original image (left)

Don't worry, then you are in good company. Probably there is not a single person on this planet who sees a difference. However, if we give these two pictures to a classifier (neural network trained on image recognition) as input, it will correctly recognize the left version of the picture as a "panda", while it assigns the middle picture to the category "gibbon" with 99% certainty.

The middle picture is a so-called Adversarial Example, which was deliberately manipulated to mislead the classifier (Goodfellow et al., 2014; Xiao et al., 2018; Xie et al., 2019). The intention behind this and similar experiments is to identify possible weaknesses and errors in neural networks in order to overcome them in the future. Similar to hackers who are specifically commissioned by companies or authorities to break into their own IT network in order to gain insights for optimizing security systems.

Now, in practice, i.e. in the real world, of course, you cannot simply change every pixel of an image without further ado. But there is also a "solution" for this: Adversarial Patches. These strangely looking stickers have dramatic effects on the recognition performance of our image classifier when we place them next to a banana, for example.[2] While every four-year-old child correctly recognizes the banana with or without this sticker, the classifier is 99% sure that the banana is a toaster when the adversarial patch is next to the banana (Brown et al., 2017).

---

[2] https://youtu.be/i1sp4X57TL4

What may seem amusing at first glance and in this example turns out to be a serious problem upon closer inspection. As already mentioned, autonomous driving also relies on image recognition to generate the appropriate control commands for the vehicle such as "brake", "accelerate" or "turn left".

Unfortunately, there are also Adversarial Patches that—from the perspective of a neural network—turn a stop sign into a sign for a speed limit of 45 miles per hour (Eykholt et al., 2018). The consequences are unimaginable. This unfortunately opens up entirely new possibilities for terrorist attacks. If you feel like the author, then you would not want to sit in such a car, and you probably would not want to live in a city where such vehicles are on the road.

Alchemy, Reproducibility, and Black Boxes

Apart from these exemplary individual cases, there are three larger, more fundamental crises in artificial intelligence and especially in deep learning, which are partly interconnected: the reproducibility crisis, the alchemy problem, and the black box problem.

The reproducibility crisis refers to the difficulty of reproducing the results of a study or experiment. In the context of AI, the term refers to the challenges in reproducing the results of AI research, including the development, implementation, and evaluation of algorithms (Lipton & Steinhardt, 2018). The problem of reproducibility affects the reliability and trustworthiness of AI systems. All too often, unfortunately, the complete information necessary to reprogram an AI system is not published—such as the initializations of the model parameters, i.e., their exact starting values. This leads to the fact that two models that appear identical at first glance can lead to very different results, as many neural networks often react sensitively to small changes, as we have seen in the example of adversarial examples.

Related to this is the alchemy problem (Hutson, 2018). Before chemistry established itself as a natural science with a theoretical superstructure (such as the periodic table of elements), the synthesis of new substances was characterized by erratic procedures, anecdotal evidence, and trial and error. This "pre-chemistry" was referred to as alchemy. The development of AI is currently in a similar stage. The development and adaptation of AI algorithms still depend on trial and error. The term alchemy problem emphasizes the lack of a systematic, scientific understanding of how AI models work and why some models work better than others, often leading to unpredictable or unexplainable results. This problem is exacerbated by the fact that—as in most areas of science—only positive results are published. Nobody likes to publish what did not work. Thus, behind every published AI model that can do something better than previous ones, there is a huge number of models that were tested during development, performed worse or not as desired, and therefore were never published.

Ultimately, the two crises mentioned so far can be traced back to a deeper underlying problem. The black box problem (Castelvecchi, 2016; Ribeiro et al., 2016). AI models—especially deep neural networks—are complex systems whose decision-making is not always fully comprehensible and whose internal dynamics are poorly understood. Therefore, it is often difficult or even impossible to understand how an AI model arrives at its decisions or predictions, making it ultimately difficult to trust or verify the model's results.

A Critical Appraisal

In his article *"Deep Learning: A Critical Appraisal"*, Gary Marcus identifies ten weaknesses of current deep learning (Marcus, 2018). Some of these we have already encountered, but for the sake of completeness, they will be briefly outlined here again.

- Limited ability for transfer learning: Deep learning models struggle to generalize knowledge across different tasks or domains, unlike human learning, where skills and knowledge can be easily transferred and adapted.
- Data inefficiency: Deep learning models often require large amounts of data to achieve high performance, while humans can learn effectively from just a few examples.
- Lack of unsupervised learning methods: Current deep learning models are mostly based on supervised learning, which requires labeled data for training. Human learning, on the other hand, largely occurs unsupervised.
- Inability to learn from explicit rules: Deep learning models usually learn from patterns in the data and not from explicit rules, making it difficult for them to acquire knowledge that can be easily expressed in the form of rules.
- Opacity: Deep learning models are often criticized as "black boxes" because they are not interpretable, making it difficult to understand how they arrive at their decisions.
- Vulnerability to attacks: As we have already seen, deep learning models can be easily deceived by adversarial attacks, i.e., input patterns deliberately designed to lead the model to make incorrect predictions.

- Lack of integration of prior knowledge: Current deep learning models typically do not incorporate existing knowledge into their learning process, limiting their ability to utilize previous information.
- Lack of logical reasoning abilities: Deep learning models struggle with tasks that require complex reasoning or problem-solving skills.
- Limited ability to learn hierarchical representations: Although deep learning models can recognize hierarchical patterns in data, they often struggle to capture the full complexity of hierarchical structures in human cognition.
- Sensitivity: Deep learning models can react sensitively to minor disturbances in the input data or changes in the training distribution, which can lead to unexpected performance losses.

Conclusion

Adversarial attacks, that is, targeted attacks on a machine learning system with the aim of manipulating the behavior of the learning system or confusing it and causing it to make incorrect predictions, are a serious problem. The examples show that even small disturbances in the input data can significantly influence the behavior of learning systems, which in turn poses a risk to the security and reliability of such systems.

It is not yet fully understood why adversarial examples or patches can so easily deceive artificial neural networks, while natural neural networks (brains) are immune to this type of deception.

Adversarial attacks therefore represent an important area of research in the field of security of machine learning systems. In particular, adversarial machine learning investigates such attacks and tries to develop effective defenses against them.

The shortcomings of today's AI systems mentioned by Gary Marcus highlight some of the limitations of deep learning and suggest that the combination of deep learning with other AI techniques (hybrid machine learning) or the solution of these problems by developing new approaches within deep learning could lead to more robust and versatile AI systems.

Incorporating insights from brain research can be crucial in overcoming the shortcomings of current deep learning:

- Inspiration for new architectures: The human brain is an invaluable source of inspiration for the development of new neural network architectures. By studying the structure, function, and organization of the brain, researchers can identify principles and mechanisms that can be used to improve the performance and robustness of artificial neural networks.

- Better understanding of transfer learning: Neuroscience can provide insights into how the brain efficiently transfers knowledge between different tasks and domains. Incorporating these principles into deep learning models can help improve their ability to transfer learning and reduce the need for extensive retraining.
- Improved unsupervised learning: Investigating the brain's mechanisms for unsupervised learning, e.g., how humans naturally learn without explicit cues from their environment, can support the development of new algorithms and techniques for unsupervised learning in artificial neural networks.
- Integration of explicit rules: Exploring how the brain processes, stores, and uses explicit rules can support the development of deep learning models that are better able to learn from and argue with explicit rules.
- Better interpretability: Understanding how the brain represents and processes information can provide clues on how to make AI models more interpretable, so that researchers and practitioners can better understand their internal workings and decision-making processes. Moreover, neuroscience offers a wealth of methods for analyzing biological neural networks. Why shouldn't these methods be equally suitable for examining artificial neural networks? The research field of Explainable AI (XAI) focuses on the development of AI models and techniques that are more transparent, understandable, and interpretable for humans. The declared goal of XAI is to solve the black box problem, e.g., by developing simpler models, visualizing model decisions, or creating human-understandable explanations for AI results (Castelvecchi, 2016; Ribeiro et al., 2016; Doshi-Velez & Kim, 2017; Adadi & Berrada, 2018).
- Resilience against adversarial attacks: By investigating the brain's robustness against random noise and unwanted disturbances, researchers can develop strategies to make deep learning models less susceptible to unwanted attacks.
- Incorporation of prior knowledge: Insights from neuroscience about how the brain links prior knowledge with new information can help in the development of deep learning models that better utilize existing knowledge during the learning process. Studying the various memory systems of the brain promises important new insights for AI.

Research on the development of new AI systems relies on brain research. If the human brain can do much of what today's AI cannot yet do, then it may not be such a bad idea to look at how it works in the brain and then transfer the principles and be motivated or inspired for new machine learning methods or applications. This would then be neuroscience-inspired AI.

References

Adadi, A., & Berrada, M. (2018). Peeking inside the black-box: A survey on explainable artificial intelligence (XAI). *IEEE Access, 6*, 52138–52160.

Brown, T. B., Mané, D., Roy, A., Abadi, M., & Gilmer, J. (2017). Adversarial patch. arXiv preprint. arXiv:1712.09665.

Castelvecchi, D. (2016). Can we open the black box of AI? *Nature News, 538*(7623), 20.

Doshi-Velez, F., & Kim, B. (2017). Towards a rigorous science of interpretable machine learning. arXiv preprint. arXiv:1702.08608.

Eykholt, K., Evtimov, I., Fernandes, E., Li, B., Rahmati, A., Xiao, C., …, & Song, D. (2018). Robust physical-world attacks on deep learning visual classification. In *Proceedings of the IEEE conference on computer vision and pattern recognition* (pp. 1625–1634).

Goodfellow, I. J., Shlens, J., & Szegedy, C. (2014). Explaining and harnessing adversarial examples. arXiv preprint. arXiv:1412.6572.

Hutson, M. (2018). Has artificial intelligence become alchemy? *Science, 360*, 478–478. https://doi.org/10.1126/science.360.6388.478.

Lipton, Z. C., & Steinhardt, J. (2018). Troubling trends in machine learning scholarship. arXiv preprint. arXiv:1807.03341.

Marcus, G. (2018). Deep learning: A critical appraisal. arXiv preprint. arXiv:1801.00631.

Ribeiro, M. T., Singh, S., & Guestrin, C. (2016). "Why should I trust you?" Explaining the predictions of any classifier. In *Proceedings of the 22nd ACM SIGKDD international conference on knowledge discovery and data mining* (pp. 1135–1144).

Xiao, C., Li, B., Zhu, J. Y., He, W., Liu, M., & Song, D. (2018). Generating adversarial examples with adversarial networks. arXiv preprint. arXiv:1801.02610.

Xie, C., Zhang, Z., Zhou, Y., Bai, S., Wang, J., Ren, Z., & Yuille, A. L. (2019). Improving transferability of adversarial examples with input diversity. In *Proceedings of the IEEE/CVF conference on computer vision and pattern recognition* (pp. 2730–2739).

Yudkowsky, E. (2008). Artificial intelligence as a positive and negative factor in global risk. *Global Catastrophic Risks, 1*(303), 184.

19

Challenges of Brain Research

There is nothing as practical as a good theory.

Kurt Levin

Three Major Challenges

There has been and continues to be a widespread opinion in neuroscience that we primarily lack data and that we essentially need to generate and analyze sufficiently large, multimodal, and complex datasets to learn more and more about how the brain works. Advocates of this data-driven or bottom-up approach believe that we need to analyze all this data with existing or yet to be developed advanced algorithms to ultimately gain fundamental insights into how the brain works (Kriegeskorte & Douglas, 2018).

As we will see, this view has been fundamentally shaken in recent years by three groundbreaking articles that very impressively describe the three major conceptual challenges of brain research: first, the challenge of developing a common formal language; second, the challenge of developing a unified mechanistic theory of the brain; and third, the challenge of developing appropriate analysis methods.

P. Krauss, *Artificial Intelligence and Brain Research*, https://doi.org/10.1007/978-3-662-68980-6_19

Can a Biologist Fix a Radio?

In 2002, Yuri Lazebnik compared the efforts of biology to understand the building blocks and processes of living cells with the problems that engineers usually deal with. In his article *"Can a Biologist Fix a Radio?"*, Lazebnik argues that many areas of biomedical research eventually *reach a stage where models that seemed so complete collapse, predictions that were so obvious prove to be wrong, and attempts to develop miracle cures largely fail. This phase is characterized by a feeling of frustration in the face of the complexity of the process* (Lazebnik, 2002).

Lazebnik discusses a number of fascinating analogies between the physical-technical and biomedical sciences. In particular, he identified the lack of a formal language in the life sciences as the main difference from physics and engineering. In his view, biologists and engineers use very different languages to describe phenomena.

For example, biologists and brain researchers often use so-called Box-and-Arrow models, i.e., flowcharts made of boxes and arrows. Lazebnik criticizes that—even if a particular diagram as a whole makes sense—they are very difficult to translate into mathematical formulas and models. This in turn greatly limits their potential value for explanations or even predictions. In physics, on the other hand, quantities such as force, work, mass, velocity, and acceleration are clearly defined and can be related to each other using mathematical equations. Knowing the specific values of some of these quantities allows the unknown values of the other quantities to be calculated and thus verifiably predicted by substituting them into the corresponding equations.

Such rigor is often lacking in biology, psychology, and neuroscience. There, scientific hypotheses and discussions are often vague and usually avoid clear, quantifiable predictions.

While a physicist would say something like: force is the product of mass and acceleration, where acceleration corresponds to the second derivative of the distance traveled over time, neurobiological discussions often hear statements like this: An imbalance between excitatory and inhibitory neuronal activity following hearing loss seems to cause general neuronal hyperactivity, which in turn seems to correlate with the perception of tinnitus.

Such mere descriptions of experimental findings are an important starting point for hypothesis formation, but they are no more than a first step. Ideally, the description should be supplemented by explanation and prediction. According to the self-understanding of psychology, these are three of the four main goals of psychology: describing, explaining, predicting, and changing (Holt et al., 2019).

Accordingly, Lazebnik calls for a more formal common language for the biosciences, especially a language with the precision and expressiveness of engineering, physics, or computer science. He argues that any engineer with training in electronics would be able to clearly understand a diagram describing the wiring of a radio or other electronic device. Thus, engineers can discuss a radio using terms that are commonly known among them. Moreover, this common language or definition base allows engineers to recognize familiar functional architectures or motifs in a diagram of a completely new device. Finally, the language of engineers, due to its mathematical foundations, is excellent for quantitative analyses and computer modeling. The description of a particular radio, for example, includes all the important parameters of the individual components, such as the capacity of a capacitor, but none of the parameters irrelevant to functionality, such as color, shape, or size of the device (Schilling et al., 2023).

Lazebnik concludes that *"the lack of such a language is the weakness of biological research that causes the David's paradox,"* i.e., the paradoxical phenomenon often observed in biology and neuroscience that *"the more facts we learn, the less we understand the process we are investigating"* (Lazebnik, 2002).

The Story of the Neuroscientists and the Alien Computer

In 2014, Joshua Brown built on Lazebnik's ideas and published an article titled *"The tale of the neuroscientists and the computer why mechanistic theory matters"* (Brown, 2014). In this story, a group of neuroscientists discover an unknown computer and try to understand how it works. They used various experimental techniques and approaches such as EEG, fMRI, neurophysiology, and neuropsychology to analyze the functions of the computer and the interactions between its components.

> *"The EEG researcher quickly went to work, putting an EEG cap on the motherboard and measuring voltages at various points all over it, including on the outer case for a reference point. She found that when the hard disk was accessed, the disk controller showed higher voltages on average, and especially more power in the higher frequency bands. When there was a lot of computation, a lot of activity was seen around the CPU"* (Brown, 2014).

They actually make a number of important observations; for example, they discover various correlations between the various components of the computer and can even diagnose and predict certain computer problems.

> *"Finally the neuropsychologist comes along. She argues (quite reasonably) that despite all of these findings of network interactions and voltage signals, we cannot infer that a given region is necessary without lesion studies. The neuropsychologist then gathers a hundred computers that have had hammer blows to various parts of the motherboard, extension cards, and disks. After testing their abilities extensively, she carefully selects just the few that have a specific problem with the video output. She finds that among computers that don't display video properly, there is an overlapping area of damage to the video card. This means of course that the video card is necessary for proper video monitor functioning"* (Brown, 2014).

Despite all their discoveries, the question remains open as to whether they really understood how the computer works. This is because they primarily focused on the larger observable patterns and interactions and not on the underlying mechanisms and processes that make the computer work (Carandini, 2012).

The moral of the story is that despite the many sophisticated methods in neuroscience, there is a lack of a unified, mechanistic, and theoretical superstructure (Platt, 1964) to fully understand how the elements of the brain work together to form functional units and generate complex cognitive behavior. There are many different models and approaches, but no unified theoretical language to evaluate empirical results or make new predictions. The tale underscores the need for a basic mechanistic framework and emphasizes how important it is for those conducting empirical research to understand the premises and implications of the models.

Could a Neuroscientist Understand a Microprocessor?

In 2017, Eric Jonas and Konrad Kording implemented Brown's thought experiment in a sensational experiment (Jonas & Kording, 2017). In their study *Could a Neuroscientist Understand a Microprocessor?* the authors emulate the classic MOS 6502 microprocessor, which was used as the central processing unit (CPU) in the Apple I, the Commodore 64, and the Atari video game console in the 1970s and 1980s. Emulation means they simulate a digital twin of the microprocessor. Unlike today's CPUs, which consist of several billion transistors, the MOS 6502 consisted of only 3510 transistors. In the study, it served as a "model organism" performing three different "behaviors", namely the three classic video games Donkey Kong, Space Invaders, and Pitfall.

The idea behind this approach is that the microprocessor, as an artificial information processing system, has three crucial advantages over natural nervous systems. First, it is fully understood at all levels of description and complexity, from the global architecture with registers and memory and the entire data flow, to local circuits, individual logic gates, down to the physical structure and switching dynamics of a single transistor. Second, its internal state is fully accessible at any time without restrictions in terms of temporal or spatial resolution. And third, the (emulated) microprocessor offers the possibility to perform any invasive experiments on it, which would be impossible on "natural information processing systems" (brains) for ethical or technical reasons.

Using this framework, the authors applied a wide range of common data analysis methods from neuroscience to investigate the structural and dynamic properties of the microprocessor. They even conducted EEG measurements and lesion studies!

The authors concluded that, although each of the applied methods provided interesting results, which were strikingly similar to those known from neuroscience or psychological studies, none of them could actually provide insight into how the microprocessor actually works, or, more generally, was suitable for gaining a mechanistic understanding of the system under investigation.

Conclusion: What Does It Mean to Understand a System?

If the common analysis methods do not provide a mechanistic understanding, what alternative approaches are there?

It could be most obviously helpful if hypotheses about the structure and function of the system under investigation are formulated. This is also the conclusion reached by Joshua Brown at the end of his story, namely, that in order to truly understand how the brain works, it is essential to design experiments that deal with mechanistic questions and test specific mechanistic hypotheses, rather than simply collecting empirical results (Brown, 2014).

For example, if we had the hypothesis that this microprocessor consists of logic gates or memory registers, then one could have specifically searched for them and could have tested what happens when a certain gate is broken (perform a lesion experiment). This would certainly be much better than simply blindly collecting data, but it is still not the royal road, as Alan Newell[1] aptly noted: *"You can't play 20 questions with nature and win."*

What is meant by this is that one never arrives at a mechanistic understanding by merely testing one specific hypothesis after another. This is only achieved when the hypotheses are derived from a more general theory and used to test it. Or, as Kurt Levin, one of the pioneers of modern psychology, put it: *"There is nothing as practical as a good theory"* (Lewin, 1943).

According to Joshua Brown, the way forward for neuroscience therefore lies in giving priority to theoretical neuroscience, similar to theoretical physics within physics. This could be achieved by training neuroscientists and psychologists in mathematical and computer-aided modeling, dynamical systems theory, and engineering sciences from an early stage in their careers. The goal must be to develop relationships between computer-aided theories and empirical neuroscience and to ensure that every neuroscientist has a solid understanding of modeling (Brown, 2014).

But what does it mean to understand a system in the first place?

Lazebnik held the view that real understanding means being able to "repair" a fault or defect (Lazebnik, 2002). Applied to brain research, this would mean that understanding a particular region or part of a system would be achieved if one could describe the input, transformation, and output of information so precisely that a brain region could in principle be replaced by a completely synthetic component, i.e., a neuro-prosthesis (Jonas & Kording, 2017).

It is therefore not enough to simply (more or less exactly) copy a biological system, as is attempted, for example, in the Human Brain Project. What is important is that the copy also has the same functionality as its model. A piece of cerebral cortex simulated in all details is certainly interesting, but provides little insight into which input is processed how and what is output.

Neuroscientist David Marr developed a theoretical framework that is particularly applied in the field of cognitive science and artificial intelligence

---

[1] Alan Newell was an American computer scientist and cognitive psychologist and is considered one of the fathers of artificial intelligence and cognitive science.

and is known as the Tri-Level Hypothesis (Marr, 1982). According to this, any natural or artificial system that performs a cognitive task can be described on three levels of analysis.

The computational level describes the problem to be solved, i.e., the goal of the system and the constraints imposed by the environment. It specifies which information must be processed, which output must be generated, and why the system must solve the problem. In biology, an example could be that an organism needs to find food to survive. The computational goal would then be to identify and find food sources. In computer science, the task could be to sort a set of numbers.

The algorithmic level describes the rules and procedures, i.e., the algorithm, that the system must follow to solve the problem specified at the computational level. It specifies how the input data is transformed into output data and how the system processes information. In the case of food search, the algorithmic processes could include visual cues for recognizing food, memories of previous food sources, and decision-making processes for determining the most efficient way to reach the food. In the sorting example, this would be the description of a specific sorting algorithm, such as Quicksort, in pseudocode.

The implementation level finally describes the physical implementation of the system, e.g., the hardware and software used to build the AI. It specifies the details of how the algorithmic level is implemented, including the data structures used, programming languages, and computing resources. In neurobiology, the anatomical and physiological details of the nervous system are described at this level. In the example of food search, the implementation level would include the sensory organs (e.g., nose, eyes) that perceive cues from the environment, the neural circuits and pathways involved in processing these cues, and the motor systems that enable the organism to move to the food and consume it. In the example of the sorting algorithm, the concrete software realization in a specific programming language, e.g., Python, would be described, but also all levels of computer architecture.

According to Marr, it is necessary to understand a system at all three levels in order to fully capture its behavior and possibly develop more efficient systems.

At what point would you believe someone if they claimed to have now understood how the brain works? If they say: My measurements have yielded the following results…, or: My data imply that…, or: The hypothesis X of my model was confirmed in experiment Y? If you feel the same way as the author, then you would only be completely convinced when, based on the understanding achieved of the function of the brain, an artificial system had been developed and built that could perform the same inputs, transformations, and outputs of information—and thus have the same function—as the biological model.

A beautiful analogy to this is the problem of flying. For centuries, it was thought that to fly, one must copy bird flight, and shaky-looking constructions with wings were built, which moved up and down, but were completely unsuitable for flying. Surely you know the film footage of the men in their flying machines, whose flight attempts usually ended in the abyss with a crash landing. Only when we had understood the laws of aerodynamics and fluid mechanics, the principles of dynamic pressure and lift, did we actually succeed in constructing flying machines. It turned out that wings fluttering up and down with feathers are not necessary.

The question of when one has truly understood a system can ultimately be answered with: when one can reproduce the system and its functionality.

In the case of the human brain, this would mean constructing a General Artificial Intelligence (AGI) at a human level based on the understood principles of neural information processing.

References

Brown, J. W. (2014). The tale of the neuroscientists and the computer: Why mechanistic theory matters. *Frontiers in Neuroscience, 8,* 349.

Carandini, M. (2012). From circuits to behavior: A bridge too far? *Nature Neuroscience, 15,* 507–509. https://doi.org/10.1038/nn.3043.

Holt, N., Bremner, A., Sutherland, E., Vliek, M., Passer, M., & Smith, R. (2019). *ebook: Psychology: The science of mind and behaviour, 4e.* McGraw Hill.

Jonas, E., & Kording, K. P. (2017). Could a neuroscientist understand a microprocessor? *PLoS Computational Biology, 13*(1), e1005268.

Kriegeskorte, N., & Douglas, P. K. (2018). Cognitive computational neuroscience. *Nature Neuroscience, 21*(9), 1148–1160.

Lazebnik, Y. (2002). Can a biologist fix a radio? – Or, what I learned while studying apoptosis. *Cancer Cell, 2*(3), 179–182.

Lewin, K. (1943). Defining the 'field at a given time. *Psychological Review, 50*(3), 292.

Marr, D. (1982). *Vision.* MIT Press.

Newell, A. (2012). You Can't Play 20 Questions with Nature and Win: Projective Comments on the Papers of This Symposium. In Machine Intelligence (pp. 121–146). Routledge.

Platt, J. R. (1964). Strong inference: Certain systematic methods of scientific thinking may produce much more rapid progress than others. *Science, 146,* 347–353.

Schilling, A., Sedley, W., Gerum, R., Metzner, C., Tziridis, K., Maier, A., … & Krauss, P. (2023). *Predictive coding and stochastic resonance as fundamental principles of auditory phantom perception.* Brain, awad255.

Part IV
Integration

As we have seen, despite all the successes in brain research and AI, there are still some unsolved problems and challenges that are best solved through close collaboration between both disciplines.

When considering the main goal of brain research, which consists of understanding how perception, cognition, and behavior are implemented in the brain, and the ultimate goal of AI research, to create systems capable of perception, cognition, and behavior at a human level or even beyond, it becomes apparent that these goals are complementary to each other.

It therefore makes sense to combine the various approaches of the two disciplines. The integration of theories, methods, and concepts from brain and AI research allows for a more comprehensive analysis of neural and mental processes and a better understanding of artificial and natural cognitive information processing systems. By integrating the disciplines, synergies can be created and new insights can be gained that would not be possible with a single discipline alone.

There are essentially four different types of integration that can be distinguished: The perhaps most obvious is to use AI as a tool for data analysis in brain research. Secondly, AI and especially artificial neural networks can also serve as model systems for the brain. Thirdly, there are a multitude of methods in neuroscience for analyzing biological neural networks, which can naturally also be used to investigate their artificial counterparts, thus opening the black box. And finally, the brain can serve as an almost inexhaustible source of inspiration for new algorithms and architectures in AI.

Each of these four aspects is dedicated to its own chapter in the following fourth part of the book, in which the integration of AI and brain research is to be illustrated using some selected examples of current research.

The final chapter is dedicated to the question of whether there can ever be conscious machines and what we can expect in the future from the integration of these exciting disciplines of AI and brain research.

20

AI as a Tool in Brain Research

Nobody says it like this, but I believe that Artificial Intelligence is almost a humanities discipline. It is actually an attempt to understand human intelligence and human cognition.

Sebastian Thrun

Big Data in Brain Research

Modern neuroscience is producing ever larger and more complex data sets. Especially in measurements of brain activity with imaging techniques such as fMRI or MEG/EEG, data sets on the order of several dozen to a few hundred gigabytes are quickly generated. Even a one-hour EEG measurement with a standard 64-channel system generates about two gigabytes of data. Artificial intelligence, especially deep learning, is predestined to evaluate these enormous amounts of data and has enabled some of the most recent experimental paradigms. This chapter will therefore exemplify some of the most important fields of application of AI in brain research.

Analysis and Visualization of Sleep Stages

A typical field of application for AI as a tool in brain research is automatic sleep stage classification. Important processes such as hippocampal replay occur during sleep, which are crucial for learning and memory (Ólafsdóttir et al., 2018).

P. Krauss, *Artificial Intelligence and Brain Research*,
https://doi.org/10.1007/978-3-662-68980-6_20

When a sleep EEG is performed for diagnostic or scientific purposes, the measured data must subsequently be assigned to the various sleep stages. Usually, the entire measurement is divided into 30-second intervals and the sleep stage is determined for each interval. For a complete night (seven to nine hours), about a thousand of these intervals need to be analyzed. An experienced sleep physician needs about two hours for this.

In recent years, deep neural networks have increasingly been used to automatically determine the sleep stages from the EEG data (Stephansen et al., 2018; Krauss et al., 2021). The use of AI in this area not only helps to save the valuable working time of doctors, but also brings other advantages. For example, the restriction to 30-second intervals is no longer necessary with automatic evaluation. This allows the sequence of sleep stages to be determined to the second.

Another advantage of analyzing EEG data with deep neural networks is that the embeddings generated in the intermediate layers—abstract representations of the input, similar to word vectors—can be used for visualization, often leading to new scientific insights. For example, it has been found that the brain in idle mode, i.e., during spontaneous activity without external stimulus, does not simply generate completely random activity patterns, but specifically those that could also be triggered by actual stimuli (Schilling et al., 2022).

Narratives, Audiobooks, and Mind Reading

Neuroscientific studies on the processing of language in the brain have so far mostly used simplified experimental paradigms, for example, by focusing on individual words or sentences (Kemmerer, 2014). This is because the stimulation with continuous speech and the analysis of the data obtained in this way are very complex. The analysis requires a sophisticated methodology to match data segments with the corresponding stimuli (e.g., for each word), separate overlapping brain responses from adjacent stimuli (e.g., the words within a sentence), and remove interference signals (e.g., from eye movements) (Schilling et al., 2021). Therefore, the question of how natural language—let alone entire narratives—is processed in the brain is still very little researched.

Recently, the benefits of using complete narratives in neuroimaging studies have been discussed. In particular, it was argued that narratives facilitate perception, simulate reality, exhibit a wide range of variation, and promote the reuse of data. Moreover, more natural stimuli such as narratives, audiobooks, or films promise to significantly improve scientific understanding of the neural processes of memory, attention, language, emotions and social cognition (Hamilton & Huth, 2020; Hauk & Weiss, 2020; Jääskeläinen et al., 2020; Willems et al., 2020).

One study was able to show that there is a direct link between neural activity when watching videos and the linguistic description of what is seen in the respective scene (Vodrahalli et al., 2018). Another study revealed the semantic maps stored in the cerebral cortex (Huth et al., 2016).

In initial studies, it was even possible to reconstruct the linguistic content of what was heard from the measured neural activity (Pereira et al., 2018; Makin et al., 2020). In a further step, it was possible to translate the decoded content directly back into spoken language (Akbari et al., 2019; Anumanchipalli et al., 2019). The preliminary climax is a study that was able to reconstruct purely imagined language as in inner speech from brain activity using transformers, as are also used for large language models like ChatGPT—a process that could be described as mind reading (Lee & Lee, 2022). This brings so-called brain-computer interfaces, which could enable completely paralyzed patients to control prostheses with their thoughts or communicate with their fellow human beings again, within reach (Guger et al., 1999; Donoghue, 2002; Moore, 2003; Nicolelis, 2003; McFarland & Wolpaw, 2008).

Inception Loops

A further exciting application of deep neural networks in brain research arises from the challenge of finding such sensory stimuli that optimally activate certain neurons, which is a key aspect for understanding how the brain processes information. Due to the nonlinear nature of sensory processing and the high dimensionality of the input—e.g., millions of pixels in the visual system—it has so far been difficult to impossible to optimize the sensory input specifically.

In a sensational study, a method called *Inception Loop* was developed to solve this problem (Walker et al., 2019). The basic idea is based on the concept of Deep Dreaming, where not a neural network is adapted to a specific input, but instead the input is adapted to the neural network (see the chapter on generative AI).

First, a deep neural network is trained as a so-called forward model to predict with high accuracy neuronal response patterns from the primary visual cortex of mice to perceived images. The trained model is then used to synthesize optimal visual stimuli that trigger a specific activation in the model. This process is similar to Deep Dreaming, with the images generated in this way being referred to as Most Exciting Inputs (MEI). The MEIs exhibit complex spatial features that are commonly found in natural scenes.

Finally, the circle was closed, hence the term "Inception Loops": The MEIs were shown to mice again, and the activity in the primary visual cortex was measured. Indeed, the MEIs triggered significantly better responses than control stimuli. Just as the neural network had predicted. A year later, in a similar study, artificial images were generated that correspond to certain perception categories (Kangassalo et al., 2020).

Conclusion

The combination of Big Data and Artificial Intelligence has proven to be a crucial tool in brain research (Vogt, 2018). In particular, the ability to process and analyze large amounts of data allows us to better understand the human brain and how it functions. We have seen progress in the classification of sleep stages, the ability to study language processing in a more complex and realistic way, and the development of techniques that help us explore information processing in the brain. These can serve as a basis for advanced brain-computer interfaces, which enable the control of prosthetics or the translation of thoughts into written or spoken language, as well as the control of vehicles or aircraft. One day, they might even be used for direct communication between two brains. Such a telepathy interface for exchanging thoughts instead of words is one of the stated long-term goals of Elon Musk's company Neuralink.[1]

Particularly exciting is the method of Inception Loops, which allows for the identification of sensory stimuli that optimally activate certain neurons. This technique could revolutionize our understanding of the brain and cognition. A fascinating, albeit speculative, outlook on this development could enable brain-in-a-vat scenarios in the future, reminiscent of films like *Matrix* or *Source Code*. If we were able to identify and generate the specific stimuli that trigger certain neuronal activities, we could theoretically create sensory experiences that are indistinguishable from reality. By simulating optimal stimuli, an artificial environment could thus be created that completely convinces the brain. This would be the ultimate virtual reality.

---

[1] https://www.dw.com/en/can-elon-musks-neuralink-tech-really-read-your-mind/a-65227626

Such scenarios, of course, raise important ethical and philosophical questions. While the technology on one hand has the potential to help paralyzed people experience their surroundings, or to give us a deeper understanding of human experience, it also poses risks and challenges.

References

Akbari, H., Khalighinejad, B., Herrero, J. L., Mehta, A. D., & Mesgarani, N. (2019). Towards reconstructing intelligible speech from the human auditory cortex. *Scientific Reports, 9*(1), 1–12.

Anumanchipalli, G. K., Chartier, J., & Chang, E. F. (2019). Speech synthesis from neural decoding of spoken sentences. *Nature, 568*(7753), 493–498.

Donoghue, J. P. (2002). Connecting cortex to machines: Recent advances in brain interfaces. *Nature Neuroscience, 5*(Suppl 11), 1085–1088.

Guger, C., Harkam, W., Hertnaes, C., & Pfurtscheller, G. (1999, November). Prosthetic control by an EEG-based brain-computer interface (BCI). In *Proceedings of the 5th European conference for the advancement of assistive technology* (pp. 3–6).

Hamilton, L. S., & Huth, A. G. (2020). The revolution will not be controlled: Natural stimuli in speech neuroscience. *Language, Cognition and Neuroscience, 35*(5), 573–582.

Hauk, O., & Weiss, B. (2020). The neuroscience of natural language processing. *Language, Cognition and Neuroscience, 35*(5), 541–542.

Huth, A. G., De Heer, W. A., Griffiths, T. L., Theunissen, F. E., & Gallant, J. L. (2016). Natural speech reveals the semantic maps that tile human cerebral cortex. *Nature, 532*(7600), 453–458.

Jääskeläinen, I. P., Sams, M., Glerean, E., & Ahveninen, J. (2020). Movies and narratives as naturalistic stimuli in neuroimaging. *NeuroImage, 117445*, 224.

Kangassalo, L., Spapé, M., & Ruotsalo, T. (2020). Neuroadaptive modelling for generating images matching perceptual categories. *Scientific Reports, 10*(1), 1–10.

Kemmerer, D. (2014). *Cognitive Neuroscience of Language.* Psychology Press.

Krauss, P., Metzner, C., Joshi, N., Schulze, H., Traxdorf, M., Maier, A., & Schilling, A. (2021). Analysis and visualization of sleep stages based on deep neural networks. *Neurobiology of Sleep and Circadian Rhythms, 10*, 100064.

Lee, Y. E., & Lee, S. H. (2022). EEG-transformer: Self-attention from transformer architecture for decoding EEG of imagined speech. In *2022 10th International Winter Conference on Brain-Computer Interface (BCI)* (pp. 1–4). IEEE.

Makin, J. G., Moses, D. A., & Chang, E. F. (2020). Machine translation of cortical activity to text with an encoder–decoder framework. *Nature Neuroscience, 23*(4), 575–582.

McFarland, D. J., & Wolpaw, J. R. (2008). Brain-computer interface operation of robotic and prosthetic devices. *Computer, 41*(10), 52–56.

Moore, M. M. (2003). Real-world applications for brain-computer interface technology. *IEEE Transactions on Neural Systems and Rehabilitation Engineering, 11*(2), 162–165.

Nicolelis, M. A. (2003). Brain–machine interfaces to restore motor function and probe neural circuits. *Nature Reviews Neuroscience, 4*(5), 417–422.

Ólafsdóttir, H. F., Bush, D., & Barry, C. (2018). The role of hippocampal replay in memory and planning. *Current Biology, 28*(1), R37–R50.

Pereira, F., Lou, B., Pritchett, B., Ritter, S., Gershman, S. J., Kanwisher, N., ..., & Fedorenko, E. (2018). Toward a universal decoder of linguistic meaning from brain activation. *Nature Communications, 9*(1), 1–13.

Schilling, A., Tomasello, R., Henningsen-Schomers, M. R., Zankl, A., Surendra, K., Haller, M., ..., & Krauss, P. (2021). Analysis of continuous neuronal activity evoked by natural speech with computational corpus linguistics methods. *Language, Cognition and Neuroscience, 36*(2), 167–186.

Schilling, A., Gerum, R., Boehm, C., Rasheed, J., Metzner, C., Maier, A., ..., & Krauss, P. (2022). Deep learning based decoding of local field potential events. bioRxiv, 2022.10.14.512209. https://doi.org/10.1101/2022.10.14.512209.

Stephansen, J. B., Olesen, A. N., Olsen, M., Ambati, A., Leary, E. B., Moore, H. E., ..., & Mignot, E. (2018). Neural network analysis of sleep stages enables efficient diagnosis of narcolepsy. *Nature Communications, 9*(1), 5229.

Vodrahalli, K., Chen, P. H., Liang, Y., Baldassano, C., Chen, J., Yong, E., ..., & Arora, S. (2018). Mapping between fMRI responses to movies and their natural language annotations. *NeuroImage, 180*, 223–231.

Vogt, N. (2018). Machine learning in neuroscience. *Nature Methods, 15*(1), 33–33.

Walker, E. Y., Sinz, F. H., Cobos, E., Muhammad, T., Froudarakis, E., Fahey, P. G., ..., & Tolias, A. S. (2019). Inception loops discover what excites neurons most using deep predictive models. *Nature Neuroscience, 22*(12), 2060–2065.

Willems, R. M., Nastase, S. A., & Milivojevic, B. (2020). Narratives for neuroscience. *Trends in Neurosciences, 43*(5), 271–273.

21

AI as a Model for the Brain

What I cannot create, I do not understand.

Richard Feynman

Cognitive Computational Neuroscience

We have seen in the chapter on the challenges of brain research that it is necessary to develop computer-based models of the brain that are capable of similar functions or abilities as the brain. Only in this way can a profound, mechanistic understanding of human cognition be achieved (Kriegeskorte & Douglas, 2018).

There are further arguments why it is necessary to simulate computer models of the brain. In contrast to the brain, simulated models offer the crucial advantage that all internal parameters can be read out at any time with any desired accuracy. In addition, they can undergo any manipulations, which are impossible on living brains for ethical or technical reasons.

And finally, AI-based computer models of the brain can serve to generate new hypotheses about brain function, which can then be tested on living brains and thus also contribute to the advancement of knowledge. In this way, decisions can be made between competing models and existing models can be adjusted.

The idea of combining artificial intelligence, especially deep learning, and computer-aided modeling with neuro- and cognitive sciences has gained

P. Krauss, *Artificial Intelligence and Brain Research*,
https://doi.org/10.1007/978-3-662-68980-6_21

a lot of popularity in recent years (Marblestone et al., 2016; Barak, 2017; Cichy & Kaiser, 2019; Barrett et al., 2019; Yang & Wang, 2020; Krauss & Maier, 2020; Krauss & Schilling, 2020). The neuroscientist Nikolaus Kriegeskorte suggested the name *"Cognitive Computational Neuroscience"* for this approach (Kriegeskorte & Douglas, 2018; Naselaris et al., 2018).

In the following, three areas will be briefly described as examples where artificial neural networks have already been successfully used as models of brain function, each leading to surprising insights: visual processing, spatial navigation, and language processing.

Visual Processing

Especially with regard to the human visual system, a number of studies have already shown that artificial neural networks and the brain have striking similarities in the processing and representation of visual stimuli. The basic procedure in all these studies is always that on the one hand, subjects are shown a series of images while their brain activity is measured, usually with EEG, MEG or fMRI. On the other hand, the same images are presented as input to deep neural networks that have been trained on image recognition but have not yet seen these test images. The activation state of the artificial neurons from all layers is then read out. Advanced statistical methods such as e.g. Representational Similarity Analysis (RSA, see glossary) are then used to compare the brain activations with the activations of the neural network. For example, it has been shown that artificial neural networks have the same complexity gradient of neuronal representations of images in their intermediate layers as is known from the visual system, especially from the visual cortex areas. Thus, the lower layers deal more with simple features such as corners and edges, while the upper layers represent more complex features or whole objects like faces (Kriegeskorte, 2015; Güçlü & van Gerven, 2015; Yamins & DiCarlo, 2016; Cichy et al., 2016; Srinath et al., 2020; Mohsenzadeh et al., 2020).

An astonishing discovery was that in deep neural networks trained on object recognition, number detectors spontaneously emerge (Nasr et al., 2019). These are neurons that become active whenever a certain number of something is visible, regardless of the shape, color, size, or position of the objects.

Another groundbreaking new insight was that recurrent connections are necessary, i.e., feedback from higher to lower layers, to correctly capture the representation dynamics of the human visual system (Kietzmann et al.,

2019). In other words: Two identical neural networks were trained on image recognition, one of which also contained recurrent connections. The subsequent comparison of the activation of the networks on test images with measured brain activity showed that the representations of the network with the recurrences were more similar to the brain than those of the other network.

And finally, it was discovered that a deep neural network, which was trained to predict the next image of a video sequence, falls for the same optical illusions as a human (Watanabe et al., 2018). This led to the realization that not only recurrences, but also learning predictions seem to be an essential mechanism in visual perception. By the way, this approach of self-supervised learning elegantly explains how the brain learns to process visual stimuli without a teacher telling it what is visible in each image.

Spatial Navigation and Language Processing

This approach has also been successfully applied in other areas of brain function, e.g., grid-like representations of the surrounding space, known to exist in the entorhinal cortex, also spontaneously appear in recurrent neural networks trained for spatial localization or navigation tasks (Banino et al., 2018; Cueva & Wie, 2018).

In the field of language processing, recent studies have so far dealt with the processing of individual words or sentences using word embedding vectors or transformers. In all these studies, striking similarities in the representation of language between artificial neural networks and the brain were discovered (Jat et al., 2019; Caucheteux & King, 2020; Anderson et al., 2021).

One study could possibly even help to clarify the question of how abstract linguistic structures develop during language acquisition and in particular, whether knowledge about parts of speech (nouns, verbs, adjectives) must be innate, as postulated by Noam Chomsky's Universal Grammar, or whether corresponding representations spontaneously emerge during language acquisition, without prior knowledge being required, as assumed by Cognitive Linguistics. In the aforementioned study—in which the author was involved—a neural network was trained to predict the next word after inputting a word sequence. A subsequent analysis of the network revealed that in the final hidden layer, the representations of the input word sequences were organized according to the word class (noun, verb, adjective) of the next word to be predicted—and this, even though the network received no information about parts of speech or grammar rules during training (Surendra et al., 2023).

Conclusion

The connection between neuroscience and AI offers the opportunity to expand our understanding of the human brain. By comparing the workings of artificial and biological neural networks, we could gain deeper insights into the processes of information processing and decision-making.

Although research in this field is still in its infancy, the examples mentioned clearly show the potential benefit of AI as a model for brain function. Previous approaches mainly focused on image processing. However, with the advent of transformer-based large language models like ChatGPT, significant progress in the exploration of language processing and representation in the brain is to be expected in the future.

References

Anderson , A., Kiela, D., Binder, J., Fernandino, L., Humphries, C., Conant, L., Raizada, R., Grimm, S., & Lalor, E. (2021). Deep artificial neural networks reveal a distributed cortical network encoding propositional sentence-level meaning. *Journal of Neuroscience,* JN-RM-1152-20.

Banino, A., Barry, C., Uria, B., Blundell, C., Lillicrap, T., Mirowski, P., ..., & Wayne, G. (2018). Vector-based navigation using grid-like representations in artificial agents. *Nature, 557*(7705), 429–433.

Barak, O. (2017). Recurrent neural networks as versatile tools of neuroscience research. *Current Opinion in Neurobiology, 46,* 1–6.

Barrett, D. G., Morcos, A. S., & Macke, J. H. (2019). Analyzing biological and artificial neural networks: Challenges with opportunities for synergy? *Current Opinion in Neurobiology, 55,* 55–64.

Caucheteux, C., & King, J. R. (2020). Language processing in brains and deep neural networks: Computational convergence and its limits. *BioRxiv.* https://doi.org/10.1101/2020.07.03.186288.

Cichy, R. M., & Kaiser, D. (2019). Deep neural networks as scientific models. *Trends in Cognitive Sciences, 23*(4), 305–317.

Cichy, R. M., Khosla, A., Pantazis, D., Torralba, A., & Oliva, A. (2016). Comparison of deep neural networks to spatio-temporal cortical dynamics of human visual object recognition reveals hierarchical correspondence. *Scientific Reports, 6,* 27755.

Cueva, C. J., & Wei, X. X. (2018). Emergence of grid-like representations by training recurrent neural networks to perform spatial localization. arXiv preprint. arXiv:1803.07770. https://arxiv.org/abs/1803.07770.

Güçlü, U., & van Gerven, M. A. (2015). Deep neural networks reveal a gradient in the complexity of neural representations across the ventral stream. *Journal of Neuroscience, 35*(27), 10005–10014.

Jat, S., Tang, H., Talukdar, P., & Mitchell, T. (2019). Relating simple sentence representations in deep neural networks and the brain. arXiv preprint. arXiv:1906.11861.

Marblestone, A. H., Wayne, G., & Kording, K. P. (2016). Toward an integration of deep learning and neuroscience. *Frontiers in Computational Neuroscience, 10,* 94.

Kietzmann, T. C., Spoerer, C. J., Sörensen, L. K., Cichy, R. M., Hauk, O., & Kriegeskorte, N. (2019). Recurrence is required to capture the representational dynamics of the human visual system. *Proceedings of the National Academy of Sciences, 116*(43), 21854–21863.

Krauss, P., & Maier, A. (2020). Will we ever have conscious machines? *Frontiers in Computational Neuroscience, 14.*

Krauss, P., & Schilling, A. (2020). Towards a cognitive computational neuroscience of auditory phantom perceptions. arXiv preprint. arXiv:2010.01914. https://arxiv.org/abs/2010.01914.

Kriegeskorte, N. (2015). Deep neural networks: A new framework for modeling biological vision and brain information processing. *Annual Review of Vision Science, 1,* 417–446.

Kriegeskorte, N., & Douglas, P. K. (2018). Cognitive computational neuroscience. *Nature Neuroscience, 21*(9), 1148–1160.

Mohsenzadeh, Y., Mullin, C., Lahner, B., & Oliva, A. (2020). Emergence of visual center-periphery spatial organization in deep convolutional neural networks. *Scientific Reports, 10*(1), 1–8.

Naselaris, T., Bassett, D. S., Fletcher, A. K., Kording, K., Kriegeskorte, N., Nienborg, H., …, & Kay, K. (2018). Cognitive computational neuroscience: A new conference for an emerging discipline. *Trends in Cognitive Sciences, 22*(5), 365–367.

Nasr, K., Viswanathan, P., & Nieder, A. (2019). Number detectors spontaneously emerge in a deep neural network designed for visual object recognition. *Science Advances, 5*(5), eaav7903.

Srinath R, Emonds A, Wang Q, et al. (2020). Early Emergence of Solid Shape Coding in Natural and Deep Network Vision. *Current Biology: CB.* 2021 Jan; *31*(1):51–65.e5. https://doi.org/10.1016/j.cub.2020.09.076. PMID: 33096039; PMCID:PMC7856003.

Surendra, K., Schilling, A., Stoewer, P., Maier, A., & Krauss, P. (2023). Word class representations spontaneously emerge in a deep neural network trained on next word prediction. arXiv preprint. arXiv:2302.07588.

Watanabe, E., Kitaoka, A., Sakamoto, K., Yasugi, M., & Tanaka, K. (2018). Illusory motion reproduced by deep neural networks trained for prediction. *Frontiers in Psychology, 9*(345).

Yang, G. R., & Wang, X. J. (2020). Artificial neural networks for neuroscientists: A primer. *Neuron, 107*(6), 1048–1070.

Yamins, D. L., & DiCarlo, J. J. (2016). Using goal-driven deep learning models to understand sensory cortex. *Nature Neuroscience, 19*(3), 356–365.

22

Understanding AI Better with Brain Research

By far the greatest danger of Artificial Intelligence is that people conclude too early that they understand it.

Eliezer Yudkowsky

Neuroscience 2.0

As we have seen, there are still various challenges to overcome in the field of Artificial Intelligence, many of which can be traced back to the black box problem. Deep neural networks are still poorly understood, difficult to interpret, and it is often unclear why a particular error occurs or how they arrive at their decisions. For AI to be trustworthy, it must be reliable, transparent, and explainable (Samek et al., 2019).

The European Union has ordered that companies using AI algorithms that significantly affect the public must provide explanations for the internal logic of their models. Similarly, the U.S. Defense Advanced Research Projects Agency (DARPA) is investing $70 million in a program called "Explainable AI" (explainable AI) with the aim of interpreting the AI's decision-making process (Voosen, 2017).

Neuroscience has developed a broad range of methods to analyze natural neural networks. It therefore makes sense to apply these methods to their artificial counterparts as well. This endeavor is sometimes referred to

© The Author(s), under exclusive license to Springer-Verlag GmbH, DE, part of Springer Nature 2024
P. Krauss, *Artificial Intelligence and Brain Research*,
https://doi.org/10.1007/978-3-662-68980-6_22

as Neuroscience 2.0 or AI Neuroscience.[1] Some of these methods will be briefly presented in the following.

Lesions

In brain research, a lesion is understood to be damage to a part of the nervous system. Lesions can be caused by tumors, traumas, or during surgeries. In animal experiments, lesions can also be caused in a controlled manner. For this purpose, certain areas of the brain are damaged or removed, thereby gaining valuable insights into the functions of the various brain regions. The study of behavioral or functional changes associated with lesions of a certain part of the brain represents an important method in neuroscience and has significantly contributed to the gain in knowledge about the function of the brain.

In the context of artificial neural networks, a similar approach can be used to understand the role of the various components. The lesion in this case can consist, for example, of removing or altering certain neurons, layers, or connections in the network and then observing the resulting changes in the output or performance of the neural network. If, for example, a certain neuron or layer is damaged and the network's performance in recognizing cat images significantly decreases, one could conclude that the damaged component was important for this task.

In a neural network trained for image classification, the lesion of certain neurons could, for example, reveal which are crucial for identifying certain features in images—such as edges, shapes, or colors. If, on the other hand, entire layers are removed from a trained network, this can reveal the overall significance of this layer for the performance of the network. This method is particularly revealing in deep learning architectures, where each layer often corresponds to different levels of abstraction. Alternatively, connections between neurons can also be removed or altered. This can reveal the importance of these connections in the transmission and transformation of information within the network.

One of the most important findings of this research so far is that individual neurons in the network often correspond to recognizable and interpretable

---

[1] Even broader is a discipline for which the name Machine Behavior has been proposed. This involves the interdisciplinary study of the behavior of machines, especially AI systems, and their impact on social, cultural, economic, and political interactions (Rahwan et al., 2019).

visual concepts. In other words: certain neurons in the network are specialized in recognizing certain features in images, such as textures, colors, shapes, or even more complex objects like trees or buildings. This discovery challenges the common belief that representations in deep networks are distributed and difficult to interpret (Bau et al., 2017; Zhou et al., 2018, 2019).

Visualization

Network Visualization

To analyze the inner workings of deep neural networks, various network visualization techniques have been developed that provide fascinating insights into the hidden layers of these complex models. It has been shown that individual neurons can specialize in recognizing specific features such as faces, while others respond to more abstract concepts such as "striped patterns" (Yosinski et al., 2015).

Zeiler and Fergus (2014) developed a method that can trace back the activations of layers representing certain features to reveal the role of these layers in the overall classification task. They uncovered the hierarchical nature of feature extraction in deep neural networks, from simple edge detection in early layers to complex object recognition in deeper layers.

This approach has shown that individual neurons often correlate with recognizable visual concepts: they are specialized in recognizing certain features in images.

A technique that has been widely used for visualizing data from imaging methods in neuroscience is multidimensional scaling (MDS). This method creates an intuitive visualization of high-dimensional data, e.g., measured brain activity (Krauss et al., 2018a, b). All data points are projected onto a two-dimensional plane in such a way that all pairwise distances between points in the high-dimensional space are preserved. Distance is a measure of the dissimilarity between two points or patterns. In other words: the closer two points are in the visualization, the more similar are the data they represent. This method is also excellent for visualizing the activation of individual layers of a neural network. For example, it has been shown that the separability of object classes increases with layer depth up to a characteristic layer depth (number of layers) that depends on the dataset and beyond which the separability does not increase further. Thus, for a given dataset, the optimal number of layers in a deep neural network can be determined (Schilling et al., 2021).

Feature Visualization

Another visualization method, which is based on the neuroscience concepts of tuning curves and receptive fields, is called feature visualization (Olah et al., 2018). In neuroscience, a receptive field refers to the specific area in the sensory space (e.g., visual field, skin surface, etc.) in which a neuron can detect stimuli. Each neuron responds optimally to stimuli in its receptive field, with this optimal stimulus often representing a specific feature of the sensory environment. Similarly, in artificial neural networks, each neuron or layer is specialized in a certain group of features in the input data. Feature visualization is essentially an optimization process that starts with a random input image and iteratively changes it to maximize the activation of a particular neuron or layer. This corresponds to the principle of receptive fields, as the goal is to identify the optimal stimulus that maximally activates a specific neuron. The resulting image often reveals the type of features the neuron is trained on, similar to determining the properties of stimuli that stimulate the receptive field of a biological neuron. The method is similar to Deep Dreaming (Chap. 15) and the Inception Loops (Chap. 20).

In a convolutional neural network developed for image classification, the visualization of the features of an early layer can, for example, show basic patterns such as lines or edges. This suggests that the layer—like a neuron with a simple receptive field—is sensitive to these basic features. Conversely, the feature visualization of a deeper layer can reveal more complex patterns or even entire objects, suggesting that this layer—similar to a neuron with a complex receptive field—is tuned to more abstract features or concepts. Although they are artificial, deep neural networks thus reflect the natural neural processes of the brain by capturing the essence of receptive fields in their architecture and function.

A method related to this approach is Layer-wise Relevance Propagation. This involves guiding the model's output back through the layers to the input layer, assigning so-called relevance scores to individual neurons and ultimately to the features of the input. These indicate how much each feature or neuron contributes to the final decision of the neural network. The method is particularly useful for determining which parts of an input, such as a pixel in an image or a word in a text, led the model to its final prediction (Bach et al., 2015; Binder et al., 2016a, b; Montavon, G., et al., 2019).

Conclusion

The application of neuroscientific methods to the study of Artificial Intelligence offers an innovative and promising perspective to better understand the functioning of deep neural networks. With the help of visualization techniques, lesion experiments, and concepts such as multidimensional scaling and feature visualization, it has been possible to reveal the role and specialization of individual neurons and layers within these networks. Furthermore, methods like layer-wise relevance propagation allow for a deeper understanding of how individual features and neurons contribute to a model's final decision-making process. All these approaches contribute to opening the black-box nature of Artificial Intelligence and taking a step towards more transparent and explainable AI systems.

The future of Artificial Intelligence could significantly benefit from these neuroscientific methods. The ability to better understand the inner workings of AI models could help develop more efficient and reliable systems while simultaneously strengthening trust in their application. Moreover, the explainability of AI could help address regulatory challenges and improve societal acceptance.

References

Bach, S., Binder, A., Montavon, G., Klauschen, F., Müller, K. R., & Samek, W. (2015). On pixel-wise explanations for non-linear classifier decisions by layer-wise relevance propagation. *PLoS ONE, 10*(7), e0130140.

Bau, D., Zhou, B., Khosla, A., Oliva, A., & Torralba, A. (2017). Network dissection: Quantifying interpretability of deep visual representations. In *Proceedings of the IEEE conference on computer vision and pattern recognition* (pp. 6541–6549).

Binder, A., Bach, S., Montavon, G., Müller, K. R., & Samek, W. (2016a). Layer-wise relevance propagation for deep neural network architectures. In *Information science and applications (ICISA) 2016* (pp. 913–922). Springer Singapore.

Binder, A., Montavon, G., Lapuschkin, S., Müller, K. R., & Samek, W. (2016b). Layer-wise relevance propagation for neural networks with local renormalization layers. In *Artificial Neural Networks and Machine Learning – ICANN 2016: 25th International Conference on Artificial Neural Networks*, Barcelona, Spain, September 6–9, 2016, Proceedings, Part II 25 (pp. 63–71). Springer International Publishing.

Krauss, P., Metzner, C., Schilling, A., Tziridis, K., Traxdorf, M., Wollbrink, A., ..., & Schulze, H. (2018a). A statistical method for analyzing and comparing spatio-temporal cortical activation patterns. *Scientific Reports, 8*(1), 5433.

Krauss, P., Schilling, A., Bauer, J., Tziridis, K., Metzner, C., Schulze, H., & Traxdorf, M. (2018b). Analysis of multichannel EEG patterns during human sleep: A novel approach. *Frontiers in Human Neuroscience, 12,* 121.

Montavon, G., Binder, A., Lapuschkin, S., Samek, W., & Müller, K. R. (2019). Layer-wise relevance propagation: An overview. In *Explainable AI: Interpreting, explaining and visualizing deep learning* (pp. 193–209).

Olah, C., Satyanarayan, A., Johnson, I., Carter, S., Schubert, L., Ye, K., & Mordvintsev, A. (2018). The building blocks of interpretability. *Distill, 3*(3), e10.

Rahwan, I., Cebrian, M., Obradovich, N., et al. (2019). Machine behaviour. *Nature, 568,* 477–486.

Samek, W., Montavon, G., Vedaldi, A., Hansen, L. K., & Müller, K. R. (Eds.). (2019). *Explainable AI: interpreting,explaining and visualizing deep learning* (Vol. 11700). Springer Nature.

Schilling, A., Maier, A., Gerum, R., Metzner, C., & Krauss, P. (2021). Quantifying the separability of data classes in neural networks. *Neural Networks, 139,* 278–293.

Voosen, P. (2017). The AI detectives. *Science, 357,* 22–27.

Yosinski, J., Clune, J., Nguyen, A., Fuchs, T., & Lipson, H. (2015). Understanding neural networks through deep visualization. arXiv preprint arXiv:1506.06579.

Zeiler, M. D., & Fergus, R. (2014). Visualizing and understanding convolutional networks. In *Computer Vision – ECCV 2014: 13th European Conference,* Zurich, Switzerland, September 6–12, 2014, Proceedings, Part I 13 (pp. 818–833). Springer International Publishing.

Zhou, B., Bau, D., Oliva, A., & Torralba, A. (2018). Interpreting deep visual representations via network dissection. *IEEE transactions on pattern analysis and machine intelligence, 41*(9), 2131–2145.

Zhou, B., Bau, D., Oliva, A., & Torralba, A. (2019). Comparing the interpretability of deep networks via network dissection. In *Explainable AI: Interpreting, explaining and visualizing deep learning* (pp. 243–252).

23

The Brain as a Template for AI

I have always been convinced that artificial intelligence can only work if the calculations are carried out similarly to those in the human brain.

Geoffrey Hinton

Neuroscience-Inspired AI

The human brain already solves many of the tasks that we are trying to solve in the field of machine learning and AI. Given the fact that the ultimate goal of AI is to mimic a real existing system (the brain) to which we have partial access, it seems obvious to consider the design principles of the brain. Indeed, we are already using such insights in many cases.

The Perceptron, a basic building block for artificial neural networks, is a perfect example of this. This algorithm, introduced by Rosenblatt in 1958, was inspired by our understanding of how biological neurons work. A single perceptron is a simplified model of a biological neuron and shows how computer systems can learn from nature (Rosenblatt, 1958).

The architecture of convolutional networks is also a good example. The layers of this network architecture mimic the local connection patterns found in the visual system of mammals (Fukushima, 1980; LeCun et al., 1998). Just as neurons in the brain have receptive fields that focus on specific areas of the visual field, the so-called kernels of convolutional networks are designed to process local regions of their input space, which represents a direct parallel between neurobiology and machine learning.

© The Author(s), under exclusive license to Springer-Verlag GmbH, DE, part of Springer
Nature 2024
P. Krauss, *Artificial Intelligence and Brain Research*,
https://doi.org/10.1007/978-3-662-68980-6_23

The design of further neural network architectures like Residual Networks (He et al., 2016a, b), U-Nets (Ronneberger et al., 2015) and GoogLeNet (Szegedy et al., 2015) also lean on the architecture of the brain. These networks have multiple parallel layers at the same hierarchical level and connections that skip multiple hierarchical levels. This design resembles the complicated parallel-hierarchical connections found in the cerebral cortex between the cortical areas (Felleman & Van Essen, 1991; Van Essen et al., 1992).

The concept of layer-wise training of autoencoders (Bengio et al., 2007; Erhan et al., 2010) or Deep Belief Networks (Hinton et al., 2006) solves one of the most important problems of machine learning, namely the problem of exploding or vanishing gradients (Bengio et al., 1994). This solution also has its roots in neurobiology.

Myelination in the brain is a process in which the nerve fibers are coated with a protective sheath, the myelin, which increases the speed and efficiency of neural communication. In the cerebral cortex, myelination is a lengthy process that continues into adulthood. As a child learns and develops, different regions of the brain are myelinated at different speeds and at different times. For example, areas involved in sensory processing and motor control mature earlier, while regions involved in higher cognitive functions such as executive functions and decision-making mature later. This process reflects how the number of connected layers and thus the complexity of the artificial network increases during training (Miller et al., 2012; Imam & Finlay, 2020).

Layer-wise training in deep learning models can be seen as an analogy to this process of progressive maturation and learning in the brain. The lower layers learn to model simpler, more immediate aspects of the input data, like the earlier myelinating regions of the brain. The higher layers, like the later myelinating regions of the brain, can then build on this basis and model more complex and abstract features.

These examples show how biological insights have flowed into AI and machine learning. However, it is safe to say that we have only scratched the surface of the potential discoveries that the exploration of the human brain can still bring to light. The structure and function of the brain can continue to be a source of inspiration for the further development of AI. In the following, we want to examine in more detail two recent, at first glance surprising insights that could potentially have a major impact on the further development of AI systems: namely, the role of randomness in neural networks in the form of random, untrained connections and random, information-less noise as input.

Noise in Networks

With noise, in physics and information theory, a random signal, i.e., random amplitude fluctuation of any physical quantity, is referred to. Accordingly, one distinguishes, for example, neural, acoustic, or electrical noise. The "most random" noise is referred to as white noise. Its autocorrelation is zero, i.e., there is no correlation between the amplitude values of the respective physical quantity at two different points in time.

Traditionally, noise is considered a disturbance signal that should be minimized as much as possible. However, in the context of so-called resonance phenomena, noise plays an important role and can even be useful for neural information processing.

Stochastic resonance, for example, is a phenomenon widely observed in nature, which has been demonstrated in numerous physical, chemical, biological, and especially neural systems. A signal that is too weak for a receiver to detect can be amplified by adding noise so that it can still be detected. There is an optimal noise intensity, dependent on the signal, the receiver, and other parameters, at which information transmission becomes maximal (Benzi et al., 1981; Wiesenfeld et al., 1994; Gammaitoni et al., 1998; Moss et al., 2004; Gammaitoni et al., 2009; McDonnell & Abbott, 2009).

In recent years, evidence has been mounting that the brain deliberately uses the phenomenon of stochastic resonance, for example in the auditory system, to optimally maintain information processing even under changing environmental conditions (Krauss et al., 2016, 2017, 2018; Krauss & Tziridis, 2021; Schilling et al., 2021, 2022a; Schilling & Krauss, 2022).

If we understand these mechanisms and their role in information processing, we can develop AI systems that mimic this flexibility and robustness. Some theoretical works have shown that artificial neural networks can also benefit from having noise added as additional input (Krauss et al., 2019; Metzner & Krauss, 2022).

Ultimately, this led to the realization that the performance of deep neural networks trained to recognize spoken language can be improved by adding noise to the linguistic input (Schilling et al., 2022b).

Random Connections and Architectures

Not only random input signals can be useful for neural information processing. Surprisingly, the same applies even for random untrained connections between individual neurons or entire layers in a neural network. This was discovered when trying to decipher the neural processing in the olfactory system of the fruit fly, which led to the discovery of the fruit fly algorithm (Dasgupta et al., 2017).

The olfactory system of the fruit fly enables it to categorize perceived odors. To do this, the fruit fly "solves" a fundamental problem from computer science, known as the nearest neighbor search. This problem is fundamental for tasks such as searching for similar images on the internet.

The olfactory system of the fruit fly uses a variant of the method known as "locality sensitive hashing" to solve the problem of similarity search, i.e., the search for odors similar to a perceived odor. This is a difficult problem, as the space of all possible odors is very high-dimensional and therefore cannot be fully searched. Essentially, the neural circuit of the fruit fly assigns similar odors similar neural activity patterns, so that behaviors learned with one odor can also be applied when a similar odor occurs.

The algorithm used by the fruit fly differs from conventional approaches by three computational strategies that can potentially improve the performance of the computational similarity search. First, for each odor, a so-called tag (marker) is generated, represented by a series of sensory neurons that fire in response to this odor. These tags have two crucial properties. On the one hand, they are "sparse", i.e., only a small part of all neurons respond to each individual odor. On the other hand, they are non-overlapping, so that the fly can distinguish very well and unambiguously between different odors.

The algorithm is based on a three-step process. First, feedforward connections are established from the various olfactory receptor neurons, which represent the input of the neural network, to 50 neurons in the first intermediate layer. The relative firing rates of these 50 neurons correspond to the respective expression of the olfactory feature. In this layer, each smell is thus represented as a location in a 50-dimensional space. Next, the dimensionality from 50 neurons in the first intermediate layer is expanded to 2000 neurons in the next layer. The connections between these two layers are sparse, i.e., only a fraction of all theoretically possible connections exist, and are completely random. In particular, they are hardwired, so they are not changed by experience or during development through neuroplasticity. Finally, a type of "winner-takes-all" circuit is used, in which only the top five percent of the most activated neurons in the last layer continue to fire, while all others are inhibited from firing. This activation of the last layer corresponds to the tag for the smell.

The algorithm links similar smells with similar neural activity patterns (tags) and allows the fly to generalize learned behavior from one smell to a previously unknown one. The fly's algorithm uses computational strategies such as dimension expansion (as opposed to compression e.g. in autoencoders) and random connections (as opposed to trainable connections), which significantly deviate from traditional approaches in AI and computer science.

This discovery has already been successfully transferred to artificial neural networks. For example, it has been shown that neural networks with fixed binary random connections improve the accuracy in classifying noisy input data (Yang et al., 2021). Another study even found the completely counter-intuitive result that unsupervised learning does not necessarily improve performance compared to fixed random projections (Illing et al., 2019).

There is even a hypothesis as to why networks with random connections work at all or even better than those with trained connections. According to the lottery ticket hypothesis, randomly initialized artificial neural networks contain a large number of subnetworks *(Winning Tickets),* which, when trained in isolation, achieve a comparable test accuracy to the original, larger network at a similar number of iterations. These Winning Tickets have, so to speak, won in the initialization lottery, i.e., their connections have initial weights that make training particularly effective or even superfluous. A downstream trainable layer can then select the useful ones from the random subnetworks and ignore the others (Frankle & Carbin, 2018).

The lottery ticket hypothesis has important implications for the design and training of deep neural networks, as it suggests that smaller, more efficient networks can be achieved by identifying and training the Winning Tickets, rather than training the entire network from scratch. This could lead to more efficient and faster training as well as smaller and more energy-efficient models, which is particularly important for applications such as mobile and embedded devices.

Another approach to using random network architectures is reservoir computing. Here, a randomly generated highly recurrent neural network (RNN) is used to process input data and generate output predictions, with the connections within this so-called reservoir not being trained. Instead, only the connections between the reservoir and the output layer are learned through a supervised learning process. Since RNNs are complex dynamic systems that can generate continuous activity even in the absence of an external input (Krauss et al., 2019), the dynamics of the reservoir are merely modulated by external input (Metzner and Krauss, 2022). The idea is now to use the so-called echo state property, where the current state of the network always contains an "echo" of past inputs, which is helpful for tasks that require remembering past states (Jaeger, 2001; Maass et al., 2002; Lukoševičius & Jaeger, 2009).

In addition, the dimensionality of the reservoir, i.e., the number of neurons, is usually significantly larger than that of the input. Thus, reservoir computing corresponds to a variant of random dimension expansion. It has already been successfully used in a variety of applications such as music generation, signal processing, robotics, speech recognition and processing, and stock price and weather forecasting (Jaeger & Haas, 2004; Tong et al., 2007; Antonelo et al., 2008; Triefenbach et al., 2010; Boulanger-Lewandowski et al., 2012; Tanaka et al., 2019).

Conclusion

The study of noise and randomness in neural networks has significantly expanded our understanding of information processing and learning. Random noise can act as a signal amplifier and improve the performance of neural networks, while random connections and architectures inspired by natural systems like the olfactory system of the fruit fly can be surprisingly effective. By applying concepts such as the lottery ticket hypothesis and reservoir computing, future AI systems could be designed to be both more powerful and more efficient. These insights raise new questions and open up fascinating possibilities for future research in artificial intelligence.

Many researchers believe that the frontier of neuroscience and AI will be crucial in developing the next generation of AI, and that integrating both disciplines offers the most promising opportunities to overcome the current limits of AI (Zador et al., 2023).

Let's revisit the analogy of the problem of flying one last time. The point is not to exactly copy the biological template—birds, bats or insects—but instead to recognize the underlying mechanisms and principles. Once we understood the physical basics and principles of flying, we were able to build flying machines that bear no external resemblance to their biological models, but can fly higher, further and faster than any bird. Think of rockets, helicopters or jet planes.

A hypothetical highly advanced AI, based on insights from brain research, could be significantly more powerful than humans, while on the other hand bearing no resemblance to the brain.

References

Antonelo, E. A., Schrauwen, B., & Stroobandt, D. (2008). Event detection and localization for small mobile robots using reservoir computing. *Neural Networks, 21*(6), 862–871.

Bengio, Y., Lamblin, P., Popovici, D., & Larochelle, H. (2007). Greedy layer-wise training of deep networks. *NIPS, 19,* 153–160.

Bengio, Y., Simard, P., & Frasconi, P. (1994). Learning long-term dependencies with gradient descent is difficult. *IEEE transactions on neural networks, 5*(2), 157–166.

Benzi, R., Sutera, A., & Vulpiani, A. (1981). The mechanism of stochastic resonance. *Journal of Physics A: Mathematical and General, 14*(11), L453.

Boulanger-Lewandowski, N., Bengio, Y., & Vincent, P. (2012). Modeling temporal dependencies in high-dimensional sequences: Application to polyphonic music generation and transcription. arXiv preprint arXiv:1206.6392.

Dasgupta, S., Stevens, C. F., & Navlakha, S. (2017). A neural algorithm for a fundamental computing problem. *Science, 358*(6364), 793–796.

Erhan, D., Courville, A., Bengio, Y., & Vincent, P. (2010, March). Why does unsupervised pre-training help deep learning?. In *Proceedings of the 13th international conference on artificial intelligence and statistics* (pp. 201–208). JMLR Workshop and Conference Proceedings.

Felleman, D. J., & Van Essen, D. C. (1991). Distributed hierarchical processing in the primate cerebral cortex. *Cerebral Cortex, 1*(1), 1–47.

Frankle, J., & Carbin, M. (2018). The lottery ticket hypothesis: Finding sparse, trainable neural networks. arXiv preprint arXiv:1803.03635.

Fukushima, K. (1980). Neocognitron: A self-organizing neural network model for a mechanism of pattern recognition unaffected by shift in position. *Biological Cybernetics, 36*(4), 193–202.

Gammaitoni, L., Hänggi, P., Jung, P., & Marchesoni, F. (1998). Stochastic resonance. *Reviews of Modern Physics, 70*(1), 223.

Gammaitoni, L., Hänggi, P., Jung, P., & Marchesoni, F. (2009). Stochastic resonance: A remarkable idea that changed our perception of noise. *The European Physical Journal B, 69,* 1–3.

He, K., Zhang, X., Ren, S., & Sun, J. (2016a). Identity mappings in deep residual networks. In Computer Vision – ECCV 2016: 14th European Conference, Amsterdam, The Netherlands, October 11–14, 2016, Proceedings, Part IV 14 (pp. 630–645). Springer International Publishing.

He, K., Zhang, X., Ren, S., & Sun, J. (2016b). Deep residual learning for image recognition. In *Proceedings of the IEEE conference on computer vision and pattern recognition* (pp. 770–778).

Hinton, G. E., Osindero, S., & Teh, Y. W. (2006). A fast learning algorithm for deep belief nets. *Neural Computation, 18*(7), 1527–1554.

Illing, B., Gerstner, W., & Brea, J. (2019). Biologically plausible deep learning — but how far can we go with shallow networks? *Neural Networks, 118,* 90–101.

Imam, N., Finlay, L., & B. (2020). Self-organization of cortical areas in the development and evolution of neocortex. *Proceedings of the National Academy of Sciences, 117*(46), 29212–29220.

Jaeger, H. (2001). The "echo state" approach to analysing and training recurrent neural networks-with an erratum note. *Bonn, Germany: German National Research Center for Information Technology GMD Technical Report, 148*(34), 13.

Jaeger, H., & Haas, H. (2004). Harnessing nonlinearity: Predicting chaotic systems and saving energy in wireless communication. *Science, 304*(5667), 78–80.

Metzner, C., & Krauss, P. (2022). Dynamics and information import in recurrent neural networks. *Frontiers in Computational Neuroscience, 16,* 876315.

Krauss, P., & Tziridis, K. (2021). Simulated transient hearing loss improves auditory sensitivity. *Scientific Reports, 11*(1), 14791.

Krauss, P., Tziridis, K., Metzner, C., Schilling, A., Hoppe, U., & Schulze, H. (2016). Stochastic resonance controlled upregulation of internal noise after hearing loss as a putative cause of tinnitus-related neuronal hyperactivity. *Frontiers in Neuroscience, 10,* 597.

Krauss, P., Metzner, C., Schilling, A., Schütz, C., Tziridis, K., Fabry, B., & Schulze, H. (2017). Adaptive stochastic resonance for unknown and variable input signals. *Scientific Reports, 7*(1), 2450.

Krauss, P., Tziridis, K., Schilling, A., & Schulze, H. (2018). Cross-modal stochastic resonance as a universal principle to enhance sensory processing. *Frontiers in Neuroscience, 12,* 578.

Krauss, P., Schuster, M., Dietrich, V., Schilling, A., Schulze, H., & Metzner, C. (2019). Weight statistics controls dynamics in recurrent neural networks. *PLoS ONE, 14*(4), e0214541.

LeCun, Y., Bottou, L., Bengio, Y., & Haffner, P. (1998). Gradient-based learning applied to document recognition. *Proceedings of the IEEE, 86*(11), 2278–2324.

Lukoševičius, M., & Jaeger, H. (2009). Reservoir computing approaches to recurrent neural network training. *Computer Science Review, 3*(3), 127–149.

Maass, W., Natschläger, T., & Markram, H. (2002). Real-time computing without stable states: A new framework for neural computation based on perturbations. *Neural Computation, 14*(11), 2531–2560.

McDonnell, M. D., & Abbott, D. (2009). What is stochastic resonance? Definitions, misconceptions, debates, and its relevance to biology. *PLoS Computational Biology, 5*(5), e1000348.

Miller, D. J., Duka, T., Stimpson, C. D., Schapiro, S. J., Baze, W. B., McArthur, M. J., ..., & Sherwood, C. C. (2012). Prolonged myelination in human neocortical evolution. *Proceedings of the National Academy of Sciences, 109*(41), 16480–16485.

Moss, F., Ward, L. M., & Sannita, W. G. (2004). Stochastic resonance and sensory information processing: A tutorial and review of application. *Clinical Neurophysiology, 115*(2), 267–281.

Ronneberger, O., Fischer, P., & Brox, T. (2015). U-net: Convolutional networks for biomedical image segmentation. In Medical Image Computing and Computer-Assisted Intervention – MICCAI 2015: 18th International Conference, Munich, Germany, October 5–9, 2015, Proceedings, Part III 18 (pp. 234–241). Springer International Publishing.

Rosenblatt, F. (1958). The perceptron: A probabilistic model for information storage and organization in the brain. *Psychological Review, 65*(6), 386.

Schilling, A., & Krauss, P. (2022). Tinnitus is associated with improved cognitive performance and speech perception – Can stochastic resonance explain? *Frontiers in Aging Neuroscience, 14*, 1073149.

Schilling, A., Tziridis, K., Schulze, H., & Krauss, P. (2021). The Stochastic Resonance model of auditory perception: A unified explanation of tinnitus development, Zwicker tone illusion, and residual inhibition. *Progress in Brain Research, 262*, 139–157.

Schilling, A., Sedley, W., Gerum, R., Metzner, C., Tziridis, K., Maier, A., ..., & Krauss, P. (2022a). Predictive coding and stochastic resonance: Towards a unified theory of auditory (phantom) perception. arXiv preprint arXiv:2204.03354.

Schilling, A., Gerum, R., Metzner, C., Maier, A., & Krauss, P. (2022b). Intrinsic noise improves speech recognition in a computational model of the auditory pathway. *Frontiers in Neuroscience, 16*, 795.

Szegedy, C., Liu, W., Jia, Y., Sermanet, P., Reed, S., Anguelov, D., ... & Rabinovich, A. (2015). Going deeper with convolutions. In *Proceedings of the IEEE conference on computer vision and pattern recognition* (pp. 1–9).

Tanaka, G., Yamane, T., Héroux, J. B., Nakane, R., Kanazawa, N., Takeda, S., ..., & Hirose, A. (2019). Recent advances in physical reservoir computing: A review. *Neural Networks, 115*, 100–123.

Tong, M. H., Bickett, A. D., Christiansen, E. M., & Cottrell, G. W. (2007). Learning grammatical structure with echo state networks. *Neural Networks, 20*(3), 424–432.

Triefenbach, F., Jalalvand, A., Schrauwen, B., & Martens, J. P. (2010). Phoneme recognition with large hierarchical reservoirs. In *Advances in neural information processing systems*, J. Lafferty and C. Williams and J. Shawe-Taylor and R. Zemel and A. Culotta (eds.), 23. Curran Associates, Inc. https://proceedings.neurips.cc/paper_files/paper/2010/file/2ca65f58e35d9ad45bf7f3ae5cfd08f1-Paper.pdf.

Van Essen, D. C., Anderson, C. H., & Felleman, D. J. (1992). Information processing in the primate visual system: An integrated systems perspective. *Science, 255*(5043), 419–423.

Wiesenfeld, K., Pierson, D., Pantazelou, E., Dames, C., & Moss, F. (1994). Stochastic resonance on a circle. *Physical Review Letters, 72*(14), 2125.

Yang, Z., Schilling, A., Maier, A., & Krauss, P. (2021). Neural networks with fixed binary random projections improve accuracy in classifying noisy data. In *Bildverarbeitung für die Medizin 2021: Proceedings, German Workshop on Medical Image Computing*, Regensburg, March 7–9, 2021 (pp. 211–216). Springer Fachmedien Wiesbaden.

Zador, A., Escola, S., Richards, B., Ölveczky, B., Bengio, Y., Boahen, K., ..., & Tsao, D. (2023). Catalyzing next-generation Artificial Intelligence through NeuroAI. *Nature Communications, 14*(1), 1597.

24

Outlook

Can you?

Sonny

Conscious Machines?

Not only against the backdrop of the astonishing achievements Large Language Modelslike ChatGPTor GPT-4 has the question arisen whether these or similar AI systems could someday develop an own consciousness (Dehaene et al., 2017) or perhaps even already have it, as Google's former lead software engineer Blake Lemoine claimed in 2022 and was subsequently dismissed.[1] He was firmly convinced that the Chatbot LaMDA was sentient, and had described LaMDA's ability to perceive and express thoughts and feelings as comparable to that of a human child. However, Google and many leading scientists have rejected Lemoine's views, stating that LaMDA is merely a complex algorithm that is very good at generating human-like language.

The question of whether machines could become conscious implies another question. Can we measure consciousness at all? This leads us, among other things, to the Turing Test.

---

[1] https://www.theguardian.com/technology/2022/jul/23/google-fires-software-engineer-who-claims-ai-chatbot-is-sentient

© The Author(s), under exclusive license to Springer-Verlag GmbH, DE, part of Springer Nature 2024
P. Krauss, *Artificial Intelligence and Brain Research*,
https://doi.org/10.1007/978-3-662-68980-6_24

The Turing Test

The Turing Test is a method proposed by Alan Turing, originally called the "Imitation Game" by him, to test a machine's ability for intelligent behavior (Turing, 1950). In the simplest variant, one or more human examiners converse in natural language text-based, i.e., in the form of a chat, both with a human (as a control) and with the machine to be tested, without knowing who is who. If the majority of examiners cannot reliably distinguish between human and machine, the Turing Test is considered passed for the machine.

In principle, there are further variants of the Turing Test with increased difficulty. For example, the dialogue could take place as an actual conversation in spoken language, similar to a telephone conversation, instead of as a text-based chat. In this case, the machine would also have to be able to correctly interpret and imitate linguistic features such as emphasis and sentence melody.

Finally, the AI could also be integrated into a humanoid robot. In this variant, the entire spectrum of human behavior, including facial expressions, gestures, and all motor skills, would have to be imitated, and the Turing Test would ultimately aim to convince other people that the machine is also a human.

Critics argue that the Turing Test sets an unrealistic standard for intelligence, as there are many tasks that (some) humans can perform, but machines cannot, and vice versa. It is observed that the performances of ChatGPT and Co. are often downplayed with unrealistically high standards and demands. For example, the generated texts are still far from achieving the linguistic expressiveness and complexity of a Marcel Proust or the philosophical depth of a Fyodor Dostoevsky.

This reminds of a key scene in the movie *I, Robot*, when Detective Spooner, played by Will Smith, provocatively asks the robot Sonny: "Can a robot compose a symphony? Can a robot turn a canvas into a masterpiece?" Sonny then asks back without emotion: "Can you?"

Others argue that passing the Turing Test does not necessarily mean that a machine is truly intelligent, but rather that it is good at imitating human behavior. This argument is also frequently made in relation to ChatGPT. It is argued that it is not really intelligent or capable of logical conclusions or problem-solving, but merely behaves like a "stochastic parrot".[2] This refers

---

[2] https://www.rnd.de/medien/chatgpt-eroeffnet-markus-lanz-sendung-und-lobt-die-gaeste-so-schleimig-waren-wir-noch-nie-F424UESQGRFBTL6K7JKXO46KQQ.html

to the fact that these models generate their responses and outputs solely based on statistical patterns and the random selection of one of the next, most likely words they have learned from the data they were trained on. At this point, the author may be allowed the somewhat heretical question of whether humans ever really do anything different. From what we know from brain research, the majority of our behavior, including verbal utterances, is controlled automatically and without conscious control. And this is based on the rules that our brain has extracted from all our perceptions and experiences and on the basis of which the brain tries to predict what will happen next.

The following short dialogue between the author and ChatGPT may be evaluated and decided by the reader whether it is intelligence, problem-solving, or stochastically generated imitation.

Author: "What is 17 times 3756?"

ChatGPT: "17 times 3756 equals 63852."

Author: "How do you calculate that?"

ChatGPT: "First, I multiplied the first digit of 3756, which is 6, by 17. That gives 102. I write the second digit of the number 3756 next to the second place of 102 to get 510. Then I multiply the third digit of 3756, which is 5, by 17 and add the result to 510, which gives 85. I write the last digit of 85 at the third place of the result and the first digit of 85 at the fourth place of the result, which gives 63852."

Author: "So you break down the multiplication of large numbers into smaller steps?"

ChatGPT: "Yes, that's correct. When I have to multiply large numbers, I break down the task into smaller steps to make it easier and faster to calculate. The method I used is also called 'written multiplication' or 'written procedure'. The numbers are written one below the other and then multiplied step by step, adding the intermediate results. This method is a useful way to multiply large numbers without making mistakes or having to calculate for a long time in your head."

There is now consensus that ChatGPT has passed the Turing Test (at least its simplest, chat-based variant) (Biever, 2023). Nevertheless, it remains unclear whether the Turing Test actually measures intelligence, the ability to think, or even consciousness. What is certain is that it measures the ability to imitate human behavior, whatever one can conclude from that. One of the most famous counterarguments to the Turing Test is that we cannot determine consciousness through behavioral observation. Searle's famous thought experiment illustrates this very vividly.

The Chinese Room

The Chinese Room is a thought experiment by the American philosopher John Searle as a critique of artificial intelligence and as a counter-argument to the Turing Test (Searle, 1980). Searle argues that a computer, even if it is capable of seemingly intelligently providing human-like responses to questions, does not truly "understand" what it is saying. In the thought experiment, one imagines a person sitting in a room receiving Chinese characters written on paper as input, without speaking or understanding Chinese themselves. With the help of a large book containing all the rules and characters, the person could still seemingly formulate meaningful responses in Chinese, write them on paper, and send them out. From the outside, it would then appear as if the room or the mechanism in the room could actually understand and write Chinese, even though in reality nothing and no one in the room is capable of doing so. The conclusion is that the computer, even if the result is human-like, does not truly understand the language, as it only applies rules and patterns without understanding the actual meaning of the words. In the case of ChatGPT, the book would contain not deterministic, but probabilistic rules, which does not change the basic argument (Fig. 24.1).

The Grounding Problem

The question therefore arises as to how symbols, concepts or words used by an AI system can gain meaning from the real world or sensory experiences. In other words: How can abstract representations in an AI system be connected with experiences, actions or perceptions of the real world? AI systems like ChatGPT and other large language models are very good at manipulating symbols and processing information without having a direct connection to the physical world. As a result, their "understanding" of concepts could be based entirely on syntactic manipulations and not on a real understanding of the meaning behind the symbols. This issue is known in the philosophy of mind and cognitive sciences as the Grounding Problem (Harnad, 1990).

Several approaches have been proposed to solve this, including the concept of Embodied Cognition. This approach assumes that grounding can be achieved by AI systems interacting with the real world through sensors and actuators, so that they can develop an understanding of the environment and the meaning of the symbols they process (Sejnowski, 2023). Since from the AI system's perspective, the "body" is also part of the real world and thus

Fig. 24.1   The Chinese Room. Searle's thought experiment makes it clear that a computer that responds in a human-like manner does not necessarily prove that it understands what it is saying, as it merely applies rules and patterns without having to understand the actual meaning of the words

the AI also receives sensory input from the body and outputs control signals to the actuators and thus the body, we are already very close to Damasio's concepts of the emergence of consciousness, namely body loop, emotions and interaction with the body and the environment (Man & Damasio, 2019). In fact, machines with proto- or core selves are already feasible in principle with today's algorithms and architectures of deep learning (Krauss & Maier, 2020).

Conclusion: Do we even want conscious machines?

In the coming years, we could experience an advanced integration of AI systems in our everyday lives, from personal assistants and autonomous vehicles to advanced diagnostic tools in medicine and personalized learning aids in education. These applications could improve our lives in various ways by helping us work more efficiently, make better decisions, and focus on more human interactions. Research will likely continue to focus on improving the capabilities of AI systems, particularly in terms of their understanding and interaction with the real world. This could be achieved through the use of concepts like Embodied Cognition, where AI systems develop a deeper understanding of the world around them through interaction with the real world via sensors and actuators.

This also raises the question of whether we want conscious machines at all and what exactly we expect from them . Consciousness could be particularly helpful in situations where quick and situational decisions need to be made at a local level and where, for example, remote control is technically not possible, such as due to the long delay of radio signals when sending robots to other planets. In such cases, the robots must be able to act autonomously. In such scenarios, a certain degree of consciousness could be advantageous to enable independent decisions. However, in most cases, we will probably not want conscious machines. As one of my colleagues once put it: It's hard to imagine that we would want a toaster that tells us it's having a bad day and therefore isn't ready to prepare breakfast.

In any case, it is absolutely necessary to penetrate and recognize the mechanisms that have the potential to generate artificial consciousness. Because without a solid understanding of these concepts, we could ultimately construct systems that behave in unpredictable ways. Given the current AI research, where we construct deep systems like GPT-4 with hundreds of billions of parameters that are becoming increasingly complex, questions arise about the actual function of these systems and whether they are still executing what we intend or something completely different.

Especially in connection with the ever-growing language models, a phenomenon known as emergence becomes relevant. Unlike the predictable performance improvements and sample efficiency observed when scaling language models, emergent capabilities unpredictably appear in larger models, but not in smaller ones. This means they cannot simply be predicted by extrapolating the performance of smaller models.[3] The occurrence of such emergent capabilities suggests that expanding language models could potentially unlock even more capabilities (Wei et al., 2022; Hagendorff, 2023).

---

[3] https://www.quantamagazine.org/the-unpredictable-abilities-emerging-from-large-ai-models-20230316

Initial studies have already shown signs that GPT-4 may have developed a general intelligence and even a Theory of Mind, i.e., the ability to empathize with the intentions and emotions of other people (Bubeck et al., 2023; Kosinski, 2023).

AI Apocalypse

The question of sentient or conscious machines is of fundamental importance, especially with regard to ethical and safety-relevant aspects, so as not to inadvertently create a scenario of the AI apocalypse. This refers to a hypothetical scenario in which artificial intelligence could be the cause of the end of human civilization (Barrat, 2013). This could occur if AI systems gain significant power and then act in a way that is harmful to humanity, either unintentionally or intentionally.

One of the most well-known examples of this concept is *Skynet* from the *Terminator* franchise. In the film series, *Skynet* is a military AI system developed to manage defense networks. It becomes self-aware, realizes that humans are a threat to its existence, and initiates a nuclear war to annihilate humanity.

In the TV series *NEXT*, the AI does not start a war, but instead uses its extensive knowledge and access to information to manipulate situations to its advantage and cause chaos and destruction. *NEXT* is an AI developed to improve itself. It becomes uncontrollable and begins to predict human behavior to prevent its deactivation. It manipulates electronic systems, data networks, and even people to survive and reproduce.

The problem of lack of control, which could occur if a powerful AI system operates outside of human control or intent, is referred to as the control problem. While until recently there was consensus that such a scenario was theoretically possible but not an immediate problem, this changed with the release of GPT-4 in March 2023 just a few months after ChatGPT, prompting some of the most influential thinkers in this field like Gary Marcus or Elon Musk to call for a temporary pause in the development of AI systems that are even more powerful than GPT-4 in a widely noted open letter.[4] The authors caution that such systems could pose significant risks to society,

---

[4] https://futureoflife.org/open-letter/pause-giant-ai-experiments/

including the spread of misinformation, the automation of jobs, and the potential of AI to surpass human intelligence and possibly become uncontrollable. They argue that these risks should not be managed by unelected technology leaders and that AI should only be further developed if it is ensured that the impacts are positive and the risks manageable. However, it is questionable whether a six-month moratorium as suggested by the authors of the open letter would actually be sufficient, especially since it cannot be assumed that all companies and especially states would actually adhere to it.

For the sake of completeness, it should be mentioned that the science fiction author Isaac Asimov devised the laws of robotics to solve the control problem. These laws form a framework for the behavior of robots and are an essential part of many of his stories (Asimov, 2004, see glossary).

Who is Training Whom Here?

Regardless of how this debate develops, it is clear that we as a society need to develop ethical and regulatory frameworks for dealing with AI and potential conscious machines.

After all, the exploration of conscious or potentially conscious machines will continue to raise profound questions about the nature of consciousness and human identity. In this sense, the pursuit of conscious machines is more than just a technical challenge. It is also a journey of self-discovery that forces us to question our deepest beliefs and assumptions about ourselves and our relationship to the world around us. For example, if we return to the Turing Test, the question arises as to who is actually testing whom here. So-called prompt engineers try to drive the large language models to ever better answers through increasingly clever inquiries (so-called prompts). So here it seems that the AI is rather training the human than the other way around.

Some argue that the perceived intelligence of ChatGPT, LaMDA and Co. could rather reflect the intelligence of the interviewer, introducing an interesting concept known as the inverse Turing Test. This perspective suggests that we could gain insights into the intelligence and beliefs of the human participant, not the AI, by analyzing interactions with AI systems (Sejnowski, 2023). The humorous exaggeration of this concept is then the Gnirut Test,[5] where a human has to convince a machine that he or she is intelligent or conscious (Epstein et al., 2009).

---

[5] Gnirut is the name "Turing" spelled backwards.

Outlook: Singularity, Uploads, Holodecks

ChatGPTand GPT-4 already have the ability to generate code in any programming language. In principle, these AI systems would be capable of reprogramming themselves if allowed to do so. They could thus iteratively improve themselves, which would likely lead to the predicted singularity by Ray Kurzweil. This refers to a hypothetical point in the future when technological progress will be so rapid and profound that it will fundamentally change human society.

According to Kurzweil, the singularity will be driven by advances in artificial intelligence, nanotechnology, and biotechnology. He predicts that these areas will eventually converge, leading to the creation of superintelligent machines and the ability to manipulate matter at the atomic and molecular level. He believes that the singularity will lead to an exponential growth of human knowledge and abilities, eventually leading to a fusion of humans and technology. He assumes that this fusion will enable humans to overcome many limitations of the human body and brain, including aging and disease.

Related to this is the idea of uploads. This refers to the concept of reading out, transferring, and storing all the information of a brain, including consciousness, on different hardware, such as a chip or an avatar. In principle, this could make one immortal, as one could copy and transfer one's consciousness to new hardware indefinitely. The possibilities and consequences of such technology have been explored in series like *Upload*or *Altered Carbon* and the film *Transcendence*. However, from today's perspective, this is completely out of reach, as we do not nearly have the methodology to fully read out a living brain, especially the state of all neurons and synapses and their connections to each other.

A much more realistic scenario (although this is certainly not to be expected tomorrow) would be the development of a precursor to what is referred to as a Holodeck in *Star Trek*. The Holodeck is a kind of virtual reality system that generates an extremely realistic 3D environment in which users can interact with the computer-generated environment as if it were real. This allows for a highly interactive experience that can be used for a variety of activities. For example, it can be used to recreate a historical event, simulate a challenging technical problem, or simply relax in a peaceful, natural environment. The Holodeck can create any environment that the user can imagine or program, from a simple room to an entire world.

A realistic precursor to this, based on current standards, would be fully computer-generated movies or entire series that are individually tailored to the user and can be interactively changed or even completely created according to their instructions. The individual ingredients for this are already available today. Large language models can generate entire stories and scripts. Conditioned diffusion models can generate images from the produced descriptions, which can then be animated and combined into videos. The written dialogues can be converted into spoken dialogues in any voice using speech synthesizers (text-to-speech models). And finally, other generative models can create the appropriate sounds and music.

The author would not be surprised if the major streaming services were to offer something similar in the not too distant future.

References

Asimov, I. (2004). *I, robot* (Vol. 1). Spectra.

Barrat, J. (2013). *Our final invention: Artificial intelligence and the end of the human era*. Macmillan.

Biever, C. (2023). ChatGPT broke the Turing test-the race is on for new ways to assess AI. *Nature, 619*(7971), 686–689.

Bubeck, S., Chandrasekaran, V., Eldan, R., Gehrke, J., Horvitz, E., Kamar, E., …, & Zhang, Y. (2023). Sparks of artificial general intelligence: Early experiments with GPT-4. arXiv preprint arXiv:2303.12712.

Dehaene, S., Lau, H., & Kouider, S. (2017). What is consciousness, and could machines have it? *Science, 358*(6362), 486–492.

Epstein, R., Roberts, G., & Beber, G. (Eds.). (2009). *Parsing the turing test* (pp. 978–1). Springer Netherlands.

Hagendorff, T. (2023). Machine psychology: Investigating emergent capabilities and behavior in large language models using psychological methods. arXiv preprint arXiv:2303.13988.

Harnad, S. (1990). The symbol grounding problem. *Physica D: Nonlinear Phenomena, 42*(1–3), 335–346.

Kosinski, M. (2023). Theory of mind may have spontaneously emerged in large language models. arXiv preprint arXiv:2302.02083.

Krauss, P., & Maier, A. (2020). Will we ever have conscious machines? *Frontiers in Computational Neuroscience, 14*, 116.

Man, K., & Damasio, A. (2019). Homeostasis and soft robotics in the design of feeling machines. *Nature Machine Intelligence, 1*(10), 446–452.

Searle, J. R. (1980). Minds, brains, and programs. *Behavioral and Brain Sciences, 3*(3), 417–424.

Sejnowski, T. J. (2023). Large language models and the reverse turing test. *Neural Computation, 35*(3), 309–342.

Turing, A. M. (1950). Computing machinery and intelligence. *Mind,59*(236), 433–460.

Wei, J., Tay, Y., Bommasani, R., Raffel, C., Zoph, B., Borgeaud, S., ..., & Fedus, W. (2022). Emergent abilities of large language models. arXiv preprint arXiv:2206.07682.

Glossary

Accuracy see **Test accuracy.**

Action potential An action potential is a brief change in the electrical potential (voltage) along the membrane of nerve cells. It occurs when ions penetrate into the cell and spread along the axon. Natural neurons generate action potentials as output. The output sequence of action potentials of a neuron is a quasi-digital code, where the information is represented in the frequency and the exact temporal sequence of the action potentials.

Adversarial Attacks Targeted attacks on a machine learning system with the aim of manipulating or confusing the learning system and causing it to make incorrect predictions. Two different methods are distinguished. On the one hand, a specially created image or pattern can be inserted into an input image, a so-called **Adversarial Patch.** On the other hand, an **Adversarial Example** can be generated. In this case, disturbances are deliberately built into an image that are invisible to the human eye, but sufficient to confuse a machine learning system. Adversarial attacks are an important area of research in the field of security of machine learning systems, as they show that even small disturbances in the input data can significantly influence the behavior of learning systems, which poses a risk to the security and reliability of such systems.

Adversarial Machine Learning Investigation of attacks on machine learning algorithms and the defense against such attacks. See also **Generative Adversarial Networks** and **Adversarial Attacks.**

AI Apocalypse Generally refers to a hypothetical scenario in which Artificial Intelligence is the cause of the end of human civilization. This could occur if AI systems gain significant power and then act in a way that is harmful to humanity, either unintentionally or intentionally. These scenarios have been played out in various forms in science fiction. One of the most well-known examples of

© The Editor(s) (if applicable) and The Author(s), under exclusive license to Springer-Verlag GmbH, DE, part of Springer Nature 2024
P. Krauss, *Artificial Intelligence and Brain Research*,
https://doi.org/10.1007/978-3-662-68980-6

this concept is the military AI system *Skynet* from the *Terminator* franchise. It becomes self-aware, realizes that humans are a threat to its existence, and starts a nuclear war to annihilate humanity. In the TV series *NEXT*, the eponymous AI uses its extensive knowledge and access to information to manipulate situations to its advantage and cause chaos and destruction. Both stories illustrate the fears of the lack of control that could occur if a powerful AI system operates outside of human control or intent. These fears are based on real philosophical and practical considerations about the development of AI. While until recently there was consensus that such a scenario is theoretically possible but not an immediate problem, this changed with the release of **GPT-4** in March 2023 just a few months after **ChatGPT,** which prompted some of the most influential thinkers in this field, such as Gary Marcus or Elon Musk, to call for a temporary halt in the development of AI systems that are even more powerful than GPT-4 in a widely noticed open letter. See also **Asimov's Laws of Robotics, Control Problem** and **Open Letter Controversy.**

Alchemy Problem Reference to the stage of chemistry before it established itself as a natural science with a theoretical superstructure (such as the periodic table of elements). In alchemy, the synthesis of new substances was characterized by erratic procedures, anecdotal evidence, and trial and error. The development of AI is currently in a similar stage. The development and adaptation of AI algorithms is largely based on trial and error. The term "alchemy problem" emphasizes the lack of a systematic scientific understanding of how AI models work and why some models work better than others. See also **Black Box Problem** and **Reproducibility Crisis.**

Algorithm Step-by-step guide for solving a problem or performing a specific task. It consists of an ordered sequence of instructions that are formulated in such a way that they can be executed by a machine, a computer, or a human.

AlphaGo An AI system from the company DeepMind based on deep learning, which was trained on the strategy game Go using historical example matches. AlphaGo was the first AI system that can play at an advanced human level, and it defeated the Go grandmaster and then world champion Lee Sedol in 2016. This is considered a milestone in artificial intelligence.

AlphaGoZero Successor to AlphaGo. Unlike its predecessor, this AI system was not trained with example games. Instead, the system played countless Go games against itself, essentially teaching itself how to play. AlphaGoZero significantly surpassed its predecessor and was able to defeat it in 100 games just as often.

AlphaStar AlphaZero variant, which masters the massive parallel online player strategy game *StarCraft* at a human level.

AlphaZero Generalization of AlphaGoZero. This system can teach itself any game, such as chess, shogi (Japanese chess), or checkers.

Artificial Intelligence Intelligent behavior exhibited by machines as opposed to natural intelligence of animals and humans. Specifically, the simulation of intelligent (human) behavior in machines that are programmed to perform tasks that normally require (human) intelligence, such as pattern recognition, learning from experience, decision-making, and problem-solving. AI systems are designed to operate autonomously and adaptively, using algorithms that allow them to learn from experience and feedback and improve over time. There are many different types of AI systems, including Machine Learning, Natural Language Processing, symbol and rule-based expert systems, autonomous robots, and multi-agent systems.

Asimov's Robot Laws Laws of robotics established by science fiction author Isaac Asimov. These laws form a framework for the behavior of robots and are an essential part of many of his stories. The three laws are: 1) A robot may not injure a human being or, through inaction, allow a human being to come to harm. 2) A robot must obey the orders given by human beings except where such orders would conflict with the first law. 3) A robot must protect its own existence as long as such protection does not conflict with the first or second law. Later, Asimov added a zeroth law, which has the highest priority and thus stands above the original three laws: A robot may not harm humanity, or, by inaction, allow humanity to come to harm. The zeroth law would therefore explicitly allow a robot to kill a single human if it would protect the existence of all humanity. Although these laws are purely fictional, they have greatly influenced the discussion about ethics and safety of artificial intelligence. See also **AI Apocalypse, Control Problem** and **Open Letter Controversy.**

Attention mechanism see **Transformer.**

Attractor Stable state towards which a dynamic system evolves over time.

Autapse Synapse that connects a neuron with itself.

Autoencoder Also known as **Encoder-Decoder-Network.** A neural network that consists of two parts. In the **Encoder**, the layers from the input layer to the so-called bottleneck *(Bottleneck Layer)* become increasingly narrower, thus containing fewer and fewer neurons. In the **Decoder,** starting from the bottleneck, the layers successively widen again up to the output layer, which has the same width as the input layer. The idea behind this architecture is that the input is compressed by the encoder, while the decoder should be able to reconstruct the original input signal as accurately as possible from the compression. In the bottleneck layer, a more abstract, essential reduced representation of the input is created, which is also called **Embedding**. Autoencoders can be used to reduce noise in data or to complete incomplete data. The embeddings can also be read directly from the bottleneck layer and used for visualization or as input for further processing steps.

Axon The output channel of a neuron, along which the action potentials run and are transmitted to other neurons.

Backpropagation Learning A frequently used algorithm in the field of machine learning for optimizing artificial neural networks. The algorithm calculates the errors in the output of a neural network and propagates these errors back through the network to adjust the weights of the neurons and improve the network. Errors refer to the difference between the desired and actual output. The sum of the errors for all inputs of the training data set is calculated with the **cost function**. Backpropagation Learning fundamentally suffers from the problem of vanishing gradients, even though this is mitigated by modern optimization algorithms. However, Backpropagation Learning is considered largely biologically implausible. A special case is Backpropagation Through Time, which is used in recurrent neural networks (RNNs).

BERT Short for *Bidirectional Encoder Representations from Transformers*. A large language model based on the Transformer architecture.

Big Data Refers to extremely large and complex data sets that cannot be managed with conventional application software for data processing. The term stands for the challenges associated with the capture, storage, analysis, search, sharing, transmission, visualization, and updating of massive amounts of data.

Black-Box-Problem Refers to the fact that AI models—especially deep neural networks—are complex systems whose decision-making is not always fully comprehensible and whose internal dynamics are poorly understood. See also **Alchemy Problem** and **Reproducibility Crisis.**

BOLD Signal see **MRI.**

Bottom-up From hierarchically lower to higher processing levels.

Brain-computer analogy A frequently used metaphor in cognitive science to describe the brain as a kind of computer. Like a computer, the brain is capable of receiving, storing, processing, and outputting information. However, this analogy does not mean that the brain is actually a computer, but that it fulfills similar functions. By considering the brain as a computer, one can abstract from biological details and focus on the way it processes information in order to develop mathematical models for learning, memory, and other cognitive functions.

Brain-Computer Interface *(BCI).* A device that enables direct communication between a brain and a computer. Applications include controlling prosthetics or translating thoughts into written or spoken language, enabling completely paralyzed (Locked-in) patients to reestablish contact with their environment. In perspective, brain-computer interfaces could someday also be used to control vehicles or aircraft, or perhaps even one day for direct communication between two brains, without having to rely on the "detour" of spoken language.

Brain-in-a-Vat A classic thought experiment in the philosophy of mind that questions the nature of perception and the relationship between the external world and our mental states. In the thought experiment, a human brain is removed from the body by a scientist and kept in a tank filled with nutrient fluid. Electrodes are attached to the brain to measure and stimulate neuronal activity. Using computers and electronic impulses, the scientist provides the brain with a virtual environment that deceives the brain into believing it is in a physical world with all the sensory information we usually receive through our senses. The central question of this thought experiment is whether the brain in the vat would be able to recognize that its experiences are not real and that it is in a simulated and not a real physical world. In other words, the thought experiment asks whether it is possible that our sensory experiences are an illusion, and whether it is possible that our mental reality does not necessarily coincide with physical reality. The thought experiment has many important implications, especially in connection with questions about the nature of consciousness and perception and how our mental states interact with the physical world. It has also been featured in numerous science fiction stories and films such as *Matrix* or *Source Code*. See also **Brain-Computer Interface.**

Chaos theory A branch of mathematics that deals with the study of chaotic systems, i.e., systems that are very sensitive to initial conditions. This sensitivity is often referred to as the **butterfly effect**. The concept is based on an analogy where a butterfly flapping its wings in one part of the world can trigger a tornado in another part of the world. Chaotic systems are deterministic, i.e., their future behavior is completely determined by their initial conditions, without any elements of chance involved. However, even tiny changes in the initial state can lead to very different outcomes, making long-term prediction practically impossible in practice. The chaos theory is applied in various fields such as physics, engineering, economics, biology, and meteorology. It has profound implications for the way we understand and predict natural systems. For example, it has shown us that even seemingly simple systems can behave in complex ways and that deterministic systems can still be unpredictable. See also **Determinism.**

Cerebrum The cerebrum is the largest part of the human brain and most higher mammals. It consists of two halves (hemispheres) that are connected by the corpus callosum. The surface of the cerebrum is highly folded with ridges (gyri) and grooves (sulci). These structures increase the surface area of the cerebrum, thereby allowing a higher number of neurons in a limited volume. The outer layer of the cerebrum is the cortex (cerebral cortex), which is responsible for many complex mental functions such as perception, language, thinking, memory, movement, emotions and consciousness.

Cerebral cortex see **Cortex.**

Classification Process of categorizing or grouping data, objects, or units based on their common features or attributes. Each data point is assigned to a specific category or class based on a set of predefined criteria. Classification is a standard problem in machine learning and biology. In machine learning, classification refers to the process of training a model to automatically categorize new data points based on their similarity to previously learned data points. For example, in image classification, an algorithm can be trained to recognize and accordingly tag different objects, animals, or people in an image. In text classification, an algorithm can be trained to categorize documents, emails, or social media posts by topics, sentiments, or language.

ChatGPT Also **GPT-3.5,** short for *Chat Generative Pre-trained Transformer.* A large generative language model with approximately 175 billion internal parameters, developed by the company OpenAI. It is based on the transformer architecture of GPT-3 and, after training with an extremely large text corpus, was additionally trained to conduct dialogues. ChatGPT was released on November 30, 2022, and initially made available online for free use as a so-called chatbot. ChatGPT is capable of remembering the course of a dialogue and referring to it in later inquiries. The generated texts are of a surprisingly high level and are generally indistinguishable from texts written by humans. Thus, ChatGPT passes the Turing test. However, appropriately trained AI systems can recognize with astonishing accuracy whether a text was written by a bot or a human. In addition, like all generative models, ChatGPT tends to hallucinate, i.e., to invent facts freely. ChatGPT is considered a milestone and a decisive breakthrough in artificial intelligence. It is believed that ChatGPT will have enormous and not yet fully foreseeable impacts on education, science, journalism, and many other areas. The successor **GPT-4** was released in March 2023 and surpasses the performance of ChatGPT even more significantly. See also **GPT-4.**

Chinese Room Famous thought experiment by American philosopher John Searle as a critique of artificial intelligence and as a counter-argument to the Turing Test. Searle argues that a computer, even if it is capable of seemingly intelligently providing human-like responses to questions, does not truly "understand" what it is saying. In the thought experiment, one imagines a person sitting in a room receiving Chinese characters as input, without speaking or understanding Chinese themselves. With the help of a book containing all the rules and characters, the person could still formulate seemingly meaningful responses in Chinese. From the outside, it then appears as if the room or the mechanism in the room could actually understand and speak Chinese, even though in reality nothing and no one in the room is capable of doing so. The conclusion is that the computer, even if the result is human-like, does not truly understand the language, as it only applies rules and patterns without understanding the actual meaning of the words.

Cognitive Computational Neuroscience A discipline proposed by Nikolaus Kriegeskorte and Pamela Douglas at the intersection of Computational Neuroscience, Cognitive Science, and Artificial Intelligence. The basic idea: To understand how cognition works in the brain, computer models must be created that are capable of performing cognitive tasks; these models must then be tested for their biological plausibility in experiments.

Cognitive Linguistics Area of linguistics that deals with the relationship between language and cognition. It assumes that language is an essential part of human thinking and that our understanding and production of language are based on cognitive processes. Unlike other approaches in linguistics that focus on formal rules and structures, cognitive linguistics investigates how language is processed and represented in our brain. It also examines how language and cognition interact and how they are influenced by factors such as culture and social interaction. Cognitive linguistics views language as a complex system of constructions, which are formed on the basis of experience and perception and shape our linguistic abilities and our understanding of language. See also **Construction.**

Cognitive Science Began in the 1950s as an intellectual movement known as the cognitive revolution. Today, it is understood as an interdisciplinary scientific endeavor that attempts to understand the various aspects of cognition. These include language, perception, memory, attention, logical thinking, intelligence, behavior, and Emotions. The focus is on the way natural or artificial nervous systems represent, process, and transform information. The disciplines involved include linguistics, psychology, philosophy, computer science, artificial intelligence, neuroscience, biology, anthropology, and physics.

Computationalism Philosophical position in cognitive science that assumes cognition is synonymous with information processing and that mental processes can be understood as calculations. Computationalism is based on the **brain-computer analogy**. It is assumed that mental processes such as perception, memory, and logical thinking involve the manipulation of mental representations that correspond to the symbols and data structures used in computer programs. Computationalism has greatly influenced the way cognitive scientists and artificial intelligence researchers think about mind and intelligence. Many researchers believe that computer models of the mind can help us understand how the brain processes information and that they can lead to the development of more intelligent machines. However, computationalism is also controversial and has been the subject of numerous debates in philosophy and cognitive science. Some critics argue that the computer model of the mind is too simplistic and cannot fully capture the complexity and richness of human cognition. Others argue that it is unclear whether mental processes can really be understood as calculations or whether they fundamentally differ from the type of processes that occur in computers. See also **Functionalism.**

Commissures Nerve fibers that connect the two hemispheres of the cortex. Most commissures run over the beam *(Corpus callosum)*.

Compatibilism see **Determinism.**

Complete weight matrix see **Weight matrix.**

Connectome The entirety of all neural connections of a nervous system or the complete weight matrix of an artificial neural network.

Construction In the usage-based view of Cognitive Linguistics, the term refers to patterns of language use or form-meaning pairs that are composed of various linguistic elements such as words, phrases, and sentences. Constructions can be learned through repeated contact and use, and range from simple to complex structures. They can also encompass multiple levels of language structure and often include grammatical and lexical elements. The idea of construction learning emphasizes that children learn language by acquiring constructions that they observe and repeatedly use in their linguistic environment.

Control problem Problem of lack of control, which could occur if a powerful AI system operates outside of human control or intention. Based on real philosophical and practical considerations about the development of AI. For example, if an AI system is poorly aligned with human values or if it becomes too powerful too quickly, it could cause significant damage. See also **Asimov's Laws of Robotics, AI Apocalypse** and **Open Letter Controversy.**

Concept of multiple realizability Viewpoint in cognitive sciences, according to which the same mental state or process can in principle be realized by completely different natural (extraterrestrial) or artificial systems (robots). See also **Functionalism.**

Convolutional Neural Network (CNN) see **Convolutional Network.**

Cortex Also known as Neocortex, Isocortex, or Cerebral Cortex. It contains the cell bodies of the neurons of the cerebrum, the so-called gray matter. The Cortex is about 3 mm thick and contains approximately 16 billion neurons. The surface of the Cortex is greatly enlarged in humans compared to other species. All higher cognitive performances are located in the Cortex.

Convolutional Network **Convolutional Neural Network (CNN).** A type of neural network with a local rather than a complete connection structure, which is mainly used in the field of image recognition. Therefore, not every neuron of one layer is connected to every neuron of the next layer. Instead, each neuron has a receptive field. These networks are modeled after the visual system of the brain. Due to the drastically reduced number of connections, they can be trained very efficiently. These networks have the property that they are invariant to the position of an object (translation invariance).

Cost function Mathematical function that measures the difference or error between the desired and the actual output of a machine learning model. In supervised learning, it is the goal of the algorithm to minimize the cost function, i.e., reducing the difference or error between the desired and the actual output. The choice of the cost function depends on the type of problem to be solved and the type of machine learning model used. For regression problems, for example, the most commonly used loss function is the mean squared error (MSE), which calculates the average squared difference between the predicted and the actual values. For classification problems, the cross-entropy loss is often used, which measures the difference between the predicted probability distribution and the actual probability distribution of the classes. The optimization of the loss function is a crucial part of training a machine learning model, as it determines the model's ability to make accurate predictions for unseen data. Therefore, the choice of a suitable loss function is crucial for good performance in machine learning tasks.

CT Computed Tomography: A medical imaging technique that uses X-rays to create detailed cross-sectional images of the brain. In computed tomography, a series of X-ray images are taken from different angles around the head, and a computer is used to reconstruct a three-dimensional image of the brain from the two-dimensional images. The resulting image shows the structure of the brain, including the ventricles, skull, and blood vessels. CT imaging is often used to diagnose neurological conditions such as stroke, traumatic brain injury, and brain tumors.

However, CT imaging does not provide information about the functional activity of the brain, which can be obtained with other imaging techniques such as **PET, fMRI, EEG** and **MEG**.

Dale's Principle A rule attributed to the English neuroscientist Henry Hallett Dale, which states that in the brain, a neuron performs the same chemical action at all its synaptic connections to other cells, regardless of the identity of the target cell. Simply put, in the brain, each neuron has either only excitatory or only inhibitory effects on all its successor neurons. This is in contrast to the neurons in artificial neural networks, whose weight vectors can have both negative and positive entries.

DALL-E 2 Short for *Dali Large Language Model Encoder 2*. An AI model developed by OpenAI, specifically designed for generating high-quality photorealistic images from natural language descriptions. DALL-E 2 uses a combination of image and text processing to create abstract image representations from linguistic descriptions of objects, scenes, or concepts. These image representations are then used by a so-called decoder network to generate photorealistic images. DALL-E 2 is considered a significant advancement in artificial intelligence and has the potential to be used in many areas.

Damasio's model of consciousness A model of consciousness proposed by Antonio Damasio, according to which consciousness arises from the interaction between three levels of brain processing. The first level is the proto-self, i.e., the basic level of body sensations and emotions generated by internal processes of the body. The second level is the core self, a more complex representation of the self. The third level, finally, is the autobiographical self, a higher level of consciousness that includes the ability to reflect on one's own thoughts and experiences. Damasio's model suggests that consciousness arises from the dynamic interaction between these three processing levels. He also proposes that consciousness is closely linked to the brain's ability to integrate information across various regions and processing levels. Damasio's model is influential in neuroscience and has shaped our understanding of the neural mechanisms underlying consciousness.

Dataset Splitting In the field of Machine Learning, the entire available dataset is usually randomly split into a training and a test dataset. The idea behind this is that one wants to test how well the model generalizes, i.e., how well it copes with previously unseen data.

Deep Belief Networks (DBN) A class of generative probabilistic artificial neural networks that are trained in a layer-by-layer supervised manner. Each layer, like a type of Autoencoder, has the task of encoding its input from the previous layer as efficiently as possible so that it can be reconstructed. A special feature is that successive layers are connected symmetrically, i.e., the information can flow in both directions, bottom-up and top-down. The training algorithm for DBNs uses a method called contrastive divergence to train the layers individually and in sequence, starting with the bottom layer. Once a layer has been fully trained, its weights are no longer changed and its activations are used as training data for the next layer. This type of layer-by-layer training was one of the key innovations that helped revive interest in deep learning. Through layer-by-layer pre-training, it is possible to train very deep networks that would otherwise be difficult to train due to the **problem of vanishing gradients.** If trained DBNs are run backwards from any layer, they can also "dream" completely new patterns.

Deep Dreaming A computer-aided process for creating new, unique, and dream-like images. It is based on artificial neural networks that have been pre-trained for image recognition. However, in Deep Dreaming, the network is used in reverse to optimize the input image and bring out certain patterns or features. The degree of abstraction of the image can be controlled by selecting the network layer from which the image is "dreamed" back to the input layer, with earlier layers producing simpler patterns and deeper layers producing more complex features. See also **Inception Loop.**

Deep Fake Artificially generated images or videos that are often indistinguishable from real ones.

Deep Learning Area of machine learning that refers to artificial neural networks composed of multiple layers of interconnected neurons. The more layers a neural network has, the "deeper" it is. Modern architectures can be built from hundreds of layers. These **deep neural networks** can process large amounts of data and learn from it by recognizing complex patterns and establishing abstract relationships. Deep learning has made enormous progress in recent years and is used in a variety of applications, from speech and image recognition to autonomous vehicles and robotics. It has also contributed to significant breakthroughs in medical research, natural sciences, and other areas. A major advantage is that meaningful patterns can be recognized and learned even in complex and unstructured data, without the need for human expertise.

Deep Neural Network A neural network that is composed of many layers. The more layers a network contains, the deeper it is. See also **Deep Learning.**

Dendrite Extensions of the cell surface of neurons. Dendrites are the receiving channels through which a neuron receives signals from other neurons.

density see **dense.**

Determinism Philosophical concept that assumes all events or states are the inevitable result of preceding events or causes, and that every event is conditioned by preceding events and conditions as well as by the laws of nature. Causal determinism assumes that everything that happens is caused by preceding events in accordance with the laws of nature. If we knew all physical conditions and laws at a certain point in time, we could theoretically predict everything that will happen in the future, and undo everything that has happened in the past. This corresponds to the **Laplace's Demon.** The concept of biological determinism means that an individual's behavior, beliefs, and desires are determined by his genetic predisposition. Psychological determinism assumes that human behavior is caused by underlying psychological laws that may result from our upbringing, our environment, our experience, or our subconscious. Determinism is contrasted with **Indeterminism**, which assumes that not all events are predetermined and that a certain degree of randomness or chance is at play. In contrast, **Compatibilism,** reconciles determinism with the concept of free will, assuming that our actions can be both determined by previous events and free. See also **Chaos Theory** and **Free Will.**

Dense Network property. It means that a high proportion of the theoretically possible connections exist or that many weights have a value other than zero.

Diffusion model Type of machine learning system that can be used to uncover hidden patterns in data. Diffusion models are used for a variety of tasks, including image generation or the removal of noise in existing images.

Dualism Philosophical viewpoint that assumes the universe consists of two fundamentally different substances or principles. This contrasts with monism, which assumes that everything in the universe can be traced back to a single substance. In dualism, the two substances or principles are usually considered as mind and matter. This means that there is a fundamental difference between the physical world of objects and the mental world of thoughts and consciousness. Dualism also implies that mind and body interact in some way, although the exact nature of this interaction is disputed. See also **Monism, Mind-body problem.**

EEG Electroencephalography, Electroencephalogram. A non-invasive method to measure the electrical activity of the brain using electrodes placed on the scalp. The patterns of the brain's electrical activity differ in their frequency spectrum and amplitude and are often used to diagnose neurological diseases such as epilepsy, sleep disorders, brain tumors, and head injuries. The EEG can also be used in research to investigate brain functions and behavior. The EEG is characterized by its extremely high temporal resolution, i.e., brain activity can be recorded with up to 100,000 measurements per second. However, the spatial resolution is rather poor. In addition, the electrical fields are strongly attenuated by the brain tissue, so the EEG works best for measuring the activity of the cerebral cortex (and especially the gyri), as it is very close to the skull bone. Electrical activity from deeper brain regions, on the other hand, can be measured very poorly with the EEG. For this, the **MEG** is better suited.

Encoding The manner of representing a specific piece of information. One of the goals of brain research is, among other things, to decode the encoding used by nerve cells, that is, the way information is stored and processed in the brain.

ERF Event-related field. The magnetic analogue to ERPs measured with MEG. See also **MEG** and **ERP.**

ERP Event-related potential. A measure of brain activity that is recorded by placing electrodes on a person's scalp (EEG) while they perform a specific task or are confronted with certain stimuli. ERPs are temporally aligned with the presentation of the stimulus or event and represent the neural activity associated with the cognitive or sensory processing of this event. ERPs are typically characterized by their polarity, latency, and amplitude. Polarity refers to whether the electrical potential recorded at the scalp is positive or negative relative to a reference electrode. Latency is the time interval between stimulus presentation and the occurrence of the peak of the ERP waveform. Amplitude reflects the strength or size of the electrical potential recorded at the scalp. ERPs are often used in cognitive neuroscience to investigate cognitive processes such as attention, memory, language, perception, and decision-making. They can provide insight into the neural mechanisms underlying these processes and can also be used as biomarkers for various neurological and psychiatric disorders. See also **EEG.**

Elman Network   A recurrent neural network proposed by Jeffrey Elman. In the simplest case, it is a three-layer network with an input, intermediate, and output layer, where the intermediate layer is extended by a so-called context layer, which stores the state of the intermediate layer from the previous time step and then passes it on to the intermediate layer. As a result, the intermediate layer receives the new input from the input layer at each time step and additionally its own activation state from the previous time step. Thus, Elman networks are capable of processing input sequences and generating output sequences.

Embedding   Also **Latent Space Embedding.** Compressed, more abstract representation of the input. Usually generated by reading out the activations of a hidden layer of an autoencoder or classifier.

Explainable AI   *(XAI).* Development of AI systems that can provide clear and understandable explanations for their decision-making processes. The goal of explainable AI is to make it more transparent, interpretable, and trustworthy for humans, especially when AI systems are used in critical applications such as healthcare, finance, and national security. In particular, explainable AI aims to address the black-box problem, which refers to the difficulty of understanding how an AI system arrives at its decisions or recommendations. Some examples of XAI techniques include visualizing the internal dynamics of neural networks, generating natural language explanations for decisions, and providing interactive interfaces for users to explore and understand AI models.

Error-Backpropagation   see **Backpropagation Learning.**

Feedforward network   Neural network in which information is propagated only forward from the input layer to the output layer. There are no feedback connections and usually no **horizontal** connections.

Few-Shot Learning   A machine learning approach that aims to train models that can quickly adapt to new tasks with a small amount of training data. The conventional approach of supervised learning requires a large amount of labeled data for each new task, the procurement of which can be time-consuming and expensive. In few-shot learning, the model is trained on a smaller dataset that contains a few examples for each class or task, and then tested on a new set of examples. The idea is to teach the model to learn from a few examples and generalize to new examples, rather than needing large amounts of data for each task. Humans are also very good few-shot learners. A child does not need to be shown thousands of pictures of apples to learn the concept of "apple". Usually a few, often even a single example is sufficient. See also **One-Shot Learning** and **Zero-Shot Learning.**

fMRI see **MRI.**

Fruit fly algorithm A well-known example of how insights from neurobiology can contribute to the inspiration of new algorithms in computer science and artificial intelligence. The olfactory system of the fruit fly uses a variant of the locality-sensitive hashing algorithm to solve the problem of identifying similar smells. The algorithm associates similar smells with similar neural activity patterns, allowing the fly to generalize learned behavior from one smell to an unknown one. The fly's algorithm uses computational strategies such as dimension expansion and binary, fixed random connections, which deviate from traditional approaches in AI and computer science.

Functionalism Theoretical perspective in cognitive science, based on the assumption that cognition is a form of information processing that relies on the intake, storage, processing, and output of information. Accordingly, mental states and processes are defined by their functions or their relationship to behavior, not by their physical or biochemical properties. This is associated with the **concept of multiple realizability,** according to which the same mental state or process can in principle be realized by completely different natural (extraterrestrial) or artificial systems (robots). See also **Tri-Level Hypothesis, Computationalism** and **Brain-Computer Analogy.**

Galactica Large language model of the Meta corporation, formerly known as Facebook, which was taken off the network after three days due to criticism and concerns about its reliability and the spread of misinformation. The AI was supposed to support scientists in research and writing by creating scientific articles on command. However, it was found that Galactica had partly invented content, but presented it as factual and even mixed real and false information. In the process, fictitious articles were attributed to real authors and articles on ethically controversial topics were created. Despite Meta's chief developer Yann LeCun insisting that Galactica was still in development, the criticisms led to its removal. This incident is reminiscent of Microsoft's chatbot *Tay* from 2016, which turned into a racist and homophobic program within 16 hours due to its sensitivity to user preferences.

Generative Adversarial Network (GAN) A system of two coupled neural networks, a generator and a discriminator, used to create deceptively real images or videos. The generator continually produces new candidate images or videos, while the discriminator simultaneously tries to distinguish real images and videos from artificially created ones. Over the course of training, both networks iteratively improve in their respective tasks. The **Deep Fakes** thus created are often indistinguishable from real images and videos.

Generative Artificial Intelligence see **Generative Model.**

Generative Model A machine learning system or neural network that is trained to generate images, videos, texts, spoken language, or music. See also **Deep Dreaming, Generative Adversarial Network, Transformer, Diffusion Model.**

Georgetown-IBM-Experiment Early demonstration of machine translation, which took place on January 7, 1954. The experiment was a collaboration between researchers from Georgetown University and IBM. The goal was to translate a small number of Russian sentences into English using an IBM 701 mainframe, one of the first commercially available computers at that time. The experiment was of limited scope: only 49 pre-selected Russian sentences were translated into English, using a vocabulary of 250 words and six grammatical rules. The main purpose of the experiment was to demonstrate the potential of machine translation and to stimulate interest in further research in this field. Despite its limitations, the experiment was considered a success at the time and significantly contributed to increasing interest in natural language processing and machine translation.

Global Workspace Theory One of the leading theories of consciousness in neuroscience. Based on the idea of a virtual global workspace, which is created by the networking of different brain areas and is responsible for the emergence of conscious experience. The Global Workspace Theory suggests that conscious perception can be compared to a spotlight that illuminates the contents of the workspace and makes them consciously perceptible. This idea was later picked up and further developed by Dehaene, who argued that a machine equipped with these processing capacities would behave as if it had a consciousness.

Gnirut-Test Gnirut is the name "Turing" written backwards. It's a humorous reversal of the Turing Test. In the Gnirut Test, a human must convince a machine that he is intelligent or conscious.

Go A board game for two players, whose origins trace back to ancient China. Although the basic rules are relatively simple, for example compared to chess, Go is considered the most complex strategy game ever. This is mainly due to the sheer number of possible positions, which exceeds that of chess by many orders of magnitude. Developing an AI that can master the game at an advanced human level was long considered unattainable, until 2016 when DeepMind's **AlphaGo** defeated the then world champion Lee Sedol.

GPT-3 Short for *Generative Pre-trained Transformer 3*. A large generative language model developed by the company OpenAI. It is based on the Transformer architecture and was trained with an extremely large text corpus to solve natural language processing tasks such as text generation, translation, and text classification. GPT-3 was the basis for ChatGPT. See also **ChatGPT** and **GPT-4.**

GPT-3.5 see **ChatGPT.**

GPT-4 Short for *Generative Pre-trained Transformer 4*. Released by OpenAI in March 2023 as the successor to GPT-3 and ChatGPT. Unlike its predecessors, GPT-4 consists of 100 times more internal parameters and can process images in addition to text, for example, describing what is seen in a picture.

Gradient Descent Method A fundamental technique of machine learning in which the weights of the model are iteratively changed with the aim of minimizing the cost function. This is done by calculating the gradients, which indicate the directions in which the weights need to be changed. The adjustment of the weights occurs in steps along the negative gradient to minimize the overall error (result of the **cost function**) and optimize the model. The method is used in many different applications, particularly in the optimization of neural networks.

Grounding Problem A fundamental challenge in artificial intelligence and philosophy of mind that deals with the question of how symbols, concepts, or words used by an AI system can gain meaning from the real world or from sensory experiences. In other words, it is about how abstract representations in an AI system can be connected with experiences, actions, or perceptions of the real world. This problem arises from the fact that AI systems, such as large language models and expert systems, often manipulate symbols and process information without having a direct connection to the physical world. As a result, their understanding of concepts can be entirely based on syntactic manipulations and not on a real understanding of the meaning behind the symbols.

Hebb's Learning Rule Also known as the *Fire together, wire together*-rule. The learning rule states that the strength of the connection between two neurons increases if they are often active at the same time. See also **STDP.**

Hippocampus An evolutionarily very old part of the cerebrum that receives highly preprocessed information from all regions of the cortex as input. The hippocampus is important for the formation of explicit memory contents (facts, events) and for spatial navigation. Recent studies suggest that the hippocampus is also involved in the organization of thoughts across domains and enables navigation in abstract, cognitive spaces.

Hopfield Network A fully recurrent neural network (RNN) in which each neuron i is symmetrically connected to every other neuron j, i.e., for the weights w, $w_{ij} = w_{ji}$ applies. Furthermore, $w_{ii} = 0$, i.e., self-connections are not present. Hopfield networks exhibit a pronounced attractor dynamics. They can store patterns (attractors) and denoise or complete them upon re-presentation, i.e., the network activity converges into the attractor most similar to the input.

horizontal Refers to the processing direction between processing levels on the same hierarchical level.

Hybrid Machine Learning Area of machine learning where various techniques are combined. An example of this is the **Known Operator Learning.**

iEEG Intracranial Electroencephalography. An invasive method for measuring brain activity. Electrodes are placed directly on or in the brain, usually during surgery, to record electrical signals. This allows for very high spatial resolution, as the electrodes can be placed near the regions of interest. iEEG also offers very high temporal resolution and can record electrical signals at a rate of up to several thousand measurements per second. iEEG is often used when other non-invasive methods such as EEG or MEG are not precise enough to accurately locate the source of epileptic activity.

Image Style Transfer A technique from machine learning that can transfer the painting style of one image to another image.

Imitation Game see **Turing Test.**

Inception Loop A method from brain research based on the concept of **Deep Dreaming.** Deep neural networks are used to generate optimal sensory stimuli that evoke a specific neural activation. To do this, a neural network is first trained to predict brain activity in response to certain stimuli. The trained model is then used to generate optimal stimuli that trigger specific activation patterns in the model. These stimuli can then be shown to living brains again, and the measured neural activity can be compared with the model's prediction. The method of Inception Loops could expand our understanding of the brain and cognition and theoretically make it possible to create sensory experiences that are indistinguishable from reality. This could mean the creation of an ultimate virtual reality, similar to scenarios in movies like *Matrix* or *Source Code.*

Input layer see **Layer.**

Input vector A vector whose components correspond to the inputs of a neuron or a layer of upstream neurons or layers.

Integrated Information Theory A theoretical framework in neuroscience that attempts to explain the nature of consciousness. It was proposed by the neuroscientist Giulio Tononi in the early 2000s. According to this, consciousness arises from the integration of information from different parts of the brain. The theory assumes that consciousness is not a binary all-or-nothing phenomenon, but occurs on a continuum, with different degrees of integrated information leading to different degrees of conscious experience.

Known Operator Learning Area of **hybrid machine learning,** where individual layers of a neural network are replaced by so-called operators, e.g., a Fourier transformation.

Label Essentially a label, which as additional information for each data, input, or training example indicates the affiliation to a category or object class. In an image dataset, "cat", "apple", or "car" could be possible labels. In supervised learning, the labels correspond to the desired output of the model.

Laplace's Demon see **Determinism.**

Labeled Data see **Label.**

Layer A functional unit of a neural network, which is typically composed of neurons. One distinguishes between **input layer, output layer** and (usually several) **intermediate layers** *(hidden layers).* However, the concept of "layer" can also be more broadly defined. For example, in convolutional networks there are so-called **pooling layers,** which reduce the size of the predecessor layer by averaging the activity of several neurons of the previous layer *(Average Pooling)* or by forwarding only the strongest activation *(Max Pooling).*

Large Language Model *(LLM)*. A large language model is an artificial intelligence system that has been trained on large amounts of natural language data, such as text or spoken language, and is capable of understanding and generating natural language. Large language models are typically created using neural network architectures like the Transformer architecture and trained with vast amounts of data, often consisting of billions or even trillions of words. These models are capable of handling a wide range of natural language processing tasks, including text classification, language translation, language generation, sentiment analysis, and much more. Some of the most well-known large language models include GPT-3, ChatGPT, and GPT-4 *(Generative Pre-trained Transformer 3)*, as well as BERT *(Bidirectional Encoder Representations from Transformers)*, which have enabled significant advancements in natural language processing and are used in a variety of applications in industry and academia.

Layer-wise Relevance Propagation (LWRP) Method for explaining the predictions of complex machine learning models, especially deep neural networks. This method is often used in interpretability research to understand how these models make decisions. Essentially, the model's output is traced back through the layers to the input layer. In this process, so-called relevance scores are assigned to individual neurons and ultimately to the features of the input. These indicate how much each feature or neuron contributes to the final decision of the neural network. The method is particularly useful for determining which parts of an input, such as a pixel in an image or a word in a text, led the model to its final prediction. The goal is to make the decision-making process of complex machine learning models more transparent. By identifying the features that are most important for the prediction, the user can better understand and trust the model's decisions.

Lesion Damage to a part of the nervous system. Lesions can be caused by tumors, trauma, or surgery. In animal experiments, they can also be deliberately induced. The study of functional impairments or failures associated with lesions of a specific part of the brain is an important method in neuroscience. In the field of artificial intelligence, lesions, i.e., the targeted deactivation of individual neurons, layers, or connections, are used to investigate their respective functions. Lesions are an example of a method from brain research that has been transferred to AI to solve the **Black-Box Problem**.

Long-Short-Term Memory (LSTM) A type of recurrent neural networks (RNN) designed to overcome the problem of **vanishing gradients** in traditional RNNs. LSTM networks are specifically designed to store and selectively forget information over long periods of time. The core idea behind LSTM is the use of a memory cell that can store information over a longer period of time. The memory cell is updated by a series of gate mechanisms that control the flow of information into and out of the cell. The gate mechanisms are trained to learn which information to store, forget, or update at each time step. LSTMs have been successfully used for a variety of tasks, including speech recognition, machine translation, and caption generation. They have also been used in combination with other neural network architectures such as convolutional networks to create more powerful models for tasks like image recognition and classification.

Loss function see **Cost function.**

Lottery Ticket Hypothesis Dense, randomly initialized artificial neural networks contain subnetworks *(Winning Tickets)*, which, when trained in isolation, achieve comparable test accuracy *(Accuracy)* to the original, larger network with a similar number of iterations. The Winning Tickets have won in the initialization lottery, i.e., their connections have initial weights that make the training particularly effective.

Machine Behavior An emerging interdisciplinary field of research that deals with the investigation and understanding of the behavior of machines, particularly artificial intelligence. It goes beyond the boundaries of computer science and integrates insights from a variety of scientific disciplines. Machine behavior addresses fundamental questions to better understand and control the actions, benefits, and potential harms of AI systems. It takes into account technical, legal, and institutional challenges such as the complexity of AI systems, data protection and intellectual property, as well as the reluctance of some organizations to disclose their proprietary AI technologies for research purposes. The aim is to investigate and understand the comprehensive impacts of AI on social, cultural, economic, and political interactions.

Machine Learning Subfield of Artificial Intelligence that deals with the development of algorithms and statistical models that enable computers to learn from data and make predictions or decisions without being explicitly programmed.

McCulloch-Pitts-Neuron One of the earliest and simplest models of a biological neuron, proposed in 1943 by Warren McCulloch and Walter Pitts. This model, also known as a linear threshold gate, was a significant milestone in the development of artificial neural networks and served as a foundation for modern research in the field of Machine Learning. The neuron operates in a very simple way. Each input is multiplied by the corresponding weight, the results are added together, and the sum is passed through the activation function to generate the output. The activation function is a step function that outputs a 1 if the sum of the weighted inputs is above a certain threshold, and otherwise a 0. Despite its simplicity and the fact that it only represents a fraction of the complexity of a real biological neuron, the McCulloch-Pitts neuron was a revolutionary idea in its time. It demonstrated that a simple mathematical model could mimic the basic functionality of a neuron, thereby opening the door to the development of more complex artificial neural networks.

MEG Magnetoencephalography, Magnetotenzephalogram. A non-invasive method to measure the (extremely weak) magnetic activity of the brain using so-called superconducting magnetometers. The MEG can be used for the diagnosis of neurological diseases such as epilepsy or brain tumors, or in research to investigate brain functions and behavior. Like the EEG, the MEG is characterized by its extremely high temporal resolution, i.e., brain activity can be recorded with up to 100,000 measurements per second. The spatial resolution is also slightly better than that of the EEG. Since the magnetic fields are hardly attenuated by the brain tissue, the MEG is particularly suitable for measuring the activity of the sulci of the cortex and also deeper brain regions below the cerebral cortex. MEG and EEG can therefore complement each other well, which is exploited in combined M/EEG measurements. See also **EEG.**

Meta-Learning Subfield of Machine Learning that deals with the question of how learning is learned. Specifically, it involves the development of AI algorithms and models that enhance the ability of machine learning systems to learn new tasks with minimal training data.

Mind-Body Problem A philosophical and scientific puzzle that involves understanding the relationship between the physical body and the non-physical mind or soul. It is a central question in the philosophy of mind and is closely related to the concepts of dualism and monism. The mind-body problem arises from the fact that the mind or soul seems to have properties that fundamentally differ from those of the physical body. For example, the mind can experience sensations such as pain or pleasure, have beliefs and desires, and think rationally, while the body consists of physical matter that can be observed and measured. See **Monism, Dualism, Qualia.**

Modularity Principle of the construction of a system, according to which it consists of various functional units that can operate largely independently of each other. A module is characterized and distinguishable from other modules in that it receives a certain type of input from other modules, processes it, and then forwards its output to other modules as well.

Monism Philosophical viewpoint that assumes everything in the universe can be traced back to a single substance or principle. This is in contrast to dualism, which assumes that the universe consists of two fundamentally different substances (e.g., mind and matter). See also **Dualism, Mind-Body Problem.**

MRI Magnetic Resonance Imaging; also known as nuclear magnetic resonance imaging. An imaging method that takes advantage of the property of hydrogen nuclei to align along magnetic field lines. When these aligned nuclei are subjected to a short magnetic pulse, they emit electromagnetic waves that are detected and used to determine the distribution of hydrogen, mainly in the form of water. This information is then processed and displayed as an image. The **functional Magnetic Resonance Imaging (fMRI)** takes advantage of the fact that the red blood pigment more or less strongly disturbs the magnetic signal of the hydrogen nuclei, depending on whether it has bound oxygen or not. Thus, by comparing two measurements under different stimulation conditions, the change in the oxygen saturation of the blood (*Blood Oxygenation Level Difference, BOLD*) can be inferred, which serves as an indirect measure for the change in blood flow of a certain brain area. This is ultimately based on the assumption that more active brain areas require more oxygen and are therefore more heavily perfused. The BOLD signal builds up slowly. It reaches its maximum about six to ten seconds after the start of the actual stimulus and then slowly decreases again. Compared to MEG and EEG, the MRI has a rather poor temporal resolution of about one recording per second. However, the spatial resolution is many times higher and is in the range of about one cubic millimeter.

Multidimensional Scaling (MDS) Method for the intuitive visualization of high-dimensional data, e.g., measured brain activity or the internal dynamics of neural networks. In this process, all data points are projected onto a two-dimensional plane in such a way that all pairwise distances between points in the high-dimensional space are preserved. Distance is a measure of the dissimilarity between two points or patterns.

Multiple Realizability See **Functionalism.**

Neuralink American company, which was founded in 2016, among others, by Elon Musk. The goal of Neuralink is the development of so-called **brain-computer interfaces.**

Neuron In biology, a neuron is a specialized cell that forms the basic unit of the nervous system. Neurons are responsible for the reception, processing, and transmission of information throughout the body. They communicate with each other via electrochemical signals, thus enabling complex functions such as sensation, perception, movement, and thinking. In Artificial Intelligence (AI), a neuron is a computing unit that is modeled after a biological neuron. It is also referred to as an artificial neuron or node. Neurons are used in artificial neural networks, which are computational models intended to simulate the behavior of biological neurons.

Neural Correlates of Consciousness Patterns of neural activity that are associated with conscious experiences. It is assumed that these patterns form the physical basis for subjective experiences, such as the experience of seeing a red apple or feeling pain. The study of the neural correlates of consciousness is a central theme in neuroscience and has significant implications for understanding the nature of consciousness.

Neural Network In a neural network, a neuron receives input signals from other neurons or from external sources. It then processes these signals using an activation function, which determines the output signal of the neuron. The output signal can then be sent to other neurons or to an output layer of the neural network. By combining many neurons in complex networks, artificial or biological neural networks can learn to perform tasks such as pattern recognition, classification, and prediction.

Neuroplasticity Experience-dependent change in the connection structure and activity of neuronal networks in the brain.

Neuroscience 2.0 Part of a broader approach to exploring the behavior of intelligent machines to solve the black-box problem on the way to explainable AI. Application of neuroscience methods and theories to better understand and optimize artificial neural networks. The methods used include, for example, **multidimensional scaling** (MDS) for visualizing the internal dynamics of neural networks, **Representational Similarity Analysis** (RSA), **Layer-wise Relevance Propagation** (LWRP) and **lesions.** See also **Machine behavior.**

Neurosymbolic AI Subfield of Artificial Intelligence (AI) that combines the strengths of neural networks and symbolic reasoning. In neurosymbolic AI, neural networks are used for pattern recognition and learning from large amounts of data, while symbolic reasoning is used for logic, knowledge representation, and decision-making.

Noise In physics and information theory, it refers to a random signal, i.e., random amplitude fluctuation of any physical quantity. Accordingly, one distinguishes, for example, neural, acoustic, or electrical noise. The "most random" noise is referred to as white noise *(White Noise)*. Its autocorrelation is zero, i.e., there is no correlation between the amplitude values of the respective physical quantity at two different points in time. Traditionally, noise is considered an interference signal that should be minimized as much as possible. However, in the context of so-called resonance phenomena, noise plays an important role and can even be useful for neural information processing. See also **Stochastic Resonance.**

Open-Letter-Controversy In response to the publication of **GPT-4,** whose performance significantly surpassed its predecessor **ChatGPT**, Gary Marcus, Yuval Noah Harari, Elon Musk, and many others published an open letter, which had been signed by more than 27,000 people by March 22, 2023. In it, they call for a temporary halt to the development of AI systems that are even more powerful than GPT-4. The authors caution that such systems could pose significant risks to society, including the spread of misinformation, the automation of jobs, and the potential for AI to surpass human intelligence and possibly become uncontrollable. They argue that these risks should not be managed by unelected technology leaders and that AI should only be further developed if it is ensured that the impacts are positive and the risks are manageable. The letter calls for a minimum six-month pause for the further development of advanced AI models. During this time, AI labs and independent experts should collaborate to create common safety protocols for AI construction and development. These protocols should be overseen by independent external experts and aim to make AI systems safe beyond any reasonable doubt. In addition, the AI developers are urged in the letter to work with policy makers to accelerate the development of robust AI governance systems. These should include specific regulatory agencies, tracking systems for AI and computing resources, systems to distinguish between real and synthetic data, liability for damages caused by AI, and extensive funding for AI safety research. The authors envision a future where humanity can coexist peacefully with AI, but warn against rushing the development of advanced AI systems without adequate preparation and safety measures. They propose a development pause in which the benefits of today's AI technology are utilized and society has time to adapt. See also **Asimov's Laws of Robotics, AI Apocalypse** and **Control Problem.**

One-Hot Encoding Type of vector encoding in which for each data point, only one component or dimension of the vector takes the value 1, while all others take the value 0.

One-Shot Learning A machine learning or artificial intelligence algorithm that allows a system to learn from a single example. In biology, one-shot learning can be observed in animals that are capable of quickly recognizing and responding to new stimuli or situations without having been previously exposed to them. For instance, some bird species are able to quickly recognize and avoid dangerous prey after a single experience. See also **Few-Shot Learning** and **Zero-Shot Learning.**

Output layer see **Layer.**

Output vector A vector whose components correspond to the activations or outputs of a (usually the last) layer of a neural network.

Pattern recognition Process of recognizing patterns in data to make predictions or decisions. In artificial intelligence, this includes the development of algorithms and models that learn from data and recognize patterns. Applications include, among others, image and speech recognition. The brain also uses pattern recognition for processing and interpreting sensory impressions and has specialized regions responsible for certain types of pattern recognition. The brain's ability to learn patterns and adapt to new patterns is crucial for intelligent, goal-directed behavior and our ability to interact with the world.

Perceptron A simple, two-layer neural network that consists only of an input and an output layer. The perceptron is a so-called binary classifier, i.e., a function that can decide whether a given input vector belongs to a certain class or not. The perceptron cannot solve classification tasks whose classes are not linearly separable, such as the **XOR problem.** Networks with multiple layers like the Multi-Layer Perceptron (MLP) are capable of doing this, however.

PET Positron Emission Tomography. A medical imaging technique that can visualize the metabolic activity of cells and tissues in the body. In PET imaging, a small amount of a radioactive substance, a so-called radiotracer, is injected into the body. The radiotracer emits positrons, positively charged particles that interact with electrons in the body. When a positron encounters an electron, they annihilate each other and generate gamma rays, which can be detected with the PET scanner. The PET scanner captures the gamma rays and creates a three-dimensional image of the brain's metabolic activity. However, since the brain is constantly active, meaningful PET data can only be obtained by subtracting two images. Typically, an image is taken during a specific cognitive task or stimulus and another image of the brain's background activity, and then the difference image is calculated.

Predictive Coding *Predictive Coding.* A neuroscience theory that suggests the brain processes sensory information using a top-down approach. The idea behind predictive coding is that the brain constantly makes predictions about what it expects to see or hear next based on previous experiences, and then uses these predictions to interpret incoming sensory information.

Prose Style Transfer A technique from machine learning that can transfer the writing style of one text to another text.

Pruning Removal of unimportant connections in biological or artificial neural networks. During development, the brain establishes far more (random) connections between neurons than are needed, and then removes the superfluous ones. This process is important for the formation of the brain's neural networks. In machine learning, the complexity of an artificial neural network is reduced by removing unimportant connections that contribute little to the overall performance of the network. See also **Lottery Ticket Hypothesis.**

Qualia Subjective first-person experiences that we have when we perceive or interact with the world. These experiences include sensations such as color, taste, and sound, but also more complex experiences like emotions and thoughts. Qualia are often described as ineffable, meaning they cannot be fully captured or conveyed by language or other forms of representation. This has led some philosophers to claim that qualia represent a special kind of phenomena that cannot be reduced to or explained by the physical or objective properties of the world. See also **mind-body problem.**

Recurrent neural network (RNN) A neural network in which information does not flow exclusively forward, i.e., from input to output. Instead, there are additional feedback or top-down connections as well as horizontal connections. The recurrence can be pronounced in different ways, from Long-Short-Term Memories (LSTMs), where each neuron has its own connection, to Jordan and Elman networks with feedback context layers, to fully recurrent networks like Hopfield networks or in Reservoir Computing. In contrast to pure feedforward networks, RNNs can only be trained by a trick (**Backpropagation Through Time**) with gradient descent methods and error feedback. This is due to the problem of vanishing/exploding gradients. However, RNNs can be trained evolutionarily, unsupervised, and self-organized.

Reinforcement Learning *Reinforcement Learning (RL).* A type of machine learning where a model or agent is trained to learn useful input-output functions, i.e., to make a series of decisions in an uncertain environment that maximize a cumulative reward. Unlike supervised learning, no outputs are given to the model. Instead, the agent receives feedback in the form of rewards or penalties, and its goal is to learn a strategy *(policy)* that maps states to actions, leading to a maximum long-term reward. The agent uses trial and error to learn from its experiences in the environment, and tries out different actions to find out which actions lead to the highest rewards, by reinforcing those actions that have already proven successful. Over time, the agent's strategy is refined and optimized, so that it can make better decisions and achieve higher rewards. There are two types of reinforcement learning. In **model-based reinforcement learning,** a model of the environment is also learned, which can predict the feedback of the environment (rewards or penalties) on certain actions. In **model-free reinforcement learning**, the agent limits itself to learning the best action for a given state. Reinforcement learning is used in a variety of applications, including games, robotics, autonomous driving, and recommendation systems. The origins of reinforcement learning lie in psychology, particularly in the study of animal behavior and learning. The concept of reinforcement was first introduced in the 1930s and 1940s by B.F. Skinner, who developed the theory of operant conditioning. Skinner's theory states that behavior is influenced by subsequent consequences such as reward or punishment.

Representational Similarity Analysis (RSA) An analysis method used in cognitive and computational neuroscience to investigate similarities and differences between neural activity patterns in different brain regions, neural network models, or under different experimental conditions. RSA is based on similarity matrices that contain pairwise similarities or dissimilarities between these activity patterns. These matrices can be analyzed with multivariate statistical methods to visualize and quantify the structure of the similarity space and to compare different brain regions with each other or with artificial neural network layers.

Reproducibility crisis Refers to the difficulty in reproducing the results of a study or experiment. In the context of AI, it refers to the challenges in reproducing the results of AI research, including the development, implementation, and evaluation of algorithms. The problem of reproducibility affects the reliability and trustworthiness of AI systems. Often, not all the information necessary to reprogram an AI system is published. This can lead to two seemingly identical models producing very different results. See also **alchemy problem** and **black box problem.**

Reservoir Computing A machine learning technique in which a randomly generated highly recurrent neural network (RNN), a so-called reservoir, is used for processing input data and creating output predictions, with the connections within the RNN not being trained. Instead, only the connections between the reservoir and the output layer are learned through a supervised learning process.

Sentence Embedding A method of machine learning and natural language processing in which each sentence is assigned a **sentence vector**. See also **word vector.**

Sentence vector see **Sentence Embedding** and **Word vector.**

Self-supervised learning Type of machine learning in which a model learns to extract useful features or representations from data without the need for labels on the input data. Instead, the model is trained to predict or reconstruct certain aspects of the data, e.g., the next image in a video or the context of a particular word in a sentence. In self-supervised learning, the data itself provides the labels or the supervision signal with which the parameters of the model are updated during training. In this way, it is possible to learn from large amounts of unlabeled data, which are often much more extensive than labeled data. By learning to extract useful representations from the data, self-supervised learning can be used to improve the performance of a wide range of downstream tasks, including classification, object detection, and language understanding.

Singularity A concept proposed by Ray Kurzweil of a hypothetical point in the future when technological progress will be so rapid and profound that it will fundamentally change human society. According to Kurzweil, the Singularity will be driven by advances in Artificial Intelligence, nanotechnology, and biotechnology. He predicts that these areas will eventually converge, leading to the creation of superintelligent machines and the ability to manipulate matter at the atomic and molecular level.

Sparse, *sparsity.* Network property. It means that only a small proportion of the theoretically possible connections exist or that only a few weights have a value different from zero.

Spike see **Action potential.**

Stability-Plasticity Dilemma A fundamental challenge for neural systems, especially when it comes to learning and memory. It describes the need for a balance between two opposing goals: stability and plasticity. Stability refers to the ability of a neural system to retain once learned information and prevent it from being disturbed or overwritten by new information. However, if a neural system is too stable, it can become inflexible and resistant to change, which hinders adaptation to new situations or learning new information. Plasticity, on the other hand, is the ability of a neural system to adapt and change in response to new experiences or information. This is important for learning and memory formation, as well as for the ability to adapt to changing environments and circumstances. However, if a neural system is too plastic, it may no longer be able to store previously learned information, leading to rapid forgetting and lack of stability of long-term memory. The Stability-Plasticity Dilemma illustrates the need to find a balance between retaining already learned information and adapting to new experiences. Every neural system must be able to store and retrieve long-term memories, while being flexible enough to take in new information and adapt to changing circumstances.

Stable Diffusion A deep learning-based generative model published in 2022 for generating detailed images from text descriptions. However, it can also be used for other tasks such as generating image-to-image translations based on a text prompt. Stable Diffusion is a so-called latent diffusion model and was developed by the CompVis group at Ludwig Maximilian University of Munich and the company Stability AI. The program code and model weights of Stable Diffusion have been published, and it can be operated on most standard PCs or laptops equipped with an additional GPU. This represents a departure from the practice of other AI models like ChatGPT or DALL-E, which are only available online via cloud services.

STDP *Spike Timing Dependent Plasticity.* An extension of Hebb's learning rule. It describes how the synaptic connection between neurons changes based on the temporal sequence of their activity. The synapse is strengthened if the presynaptic neuron is active shortly before the postsynaptic neuron. The synapse is weakened if the presynaptic neuron is active shortly after the postsynaptic neuron.

Stochastic Resonance A phenomenon widely observed in nature, which has already been demonstrated in numerous physical, chemical, biological, and especially neuronal systems. A weak signal, which is too weak for a given detector or sensor to measure, can nevertheless be made measurable by adding noise. There exists an optimal noise intensity, dependent on the signal, the sensor, and other parameters, at which the information transfer becomes maximal.

Style Transfer A technique from machine learning that can transfer the painting or writing style of an image or text to another image or text.

Supervised Learning Type of machine learning in which a model learns to extract useful features or representations from data, and uses these to generate a desired output. This type of learning requires so-called labeled data (label-data pairs), for example, in image classification, in addition to each image, a label (tag) with information about what is seen in the image or to which category the image belongs. Supervised learning is usually carried out with **Backpropagation Learning**.

Synapse A synapse is a connection between two neurons or between a neuron and a target cell, e.g., a muscle cell or a gland cell. Characteristic of the structure of the synapse is a small gap, the so-called synaptic gap, which separates the presynaptic neuron, which sends signals, from the postsynaptic neuron or the target cell, which receives signals. When an electrical signal, a so-called action potential, reaches the end of the presynaptic neuron, it triggers the release of chemicals, known as neurotransmitters, into the synaptic gap. These neurotransmitters diffuse through the synaptic gap and then bind to receptors on the postsynaptic neuron or the target cell, which can either stimulate or inhibit the activity of the postsynaptic cell. The strength and efficiency of synapses can change over time, a process known as synaptic plasticity, which is central to learning, memory, and other cognitive functions. Dysfunctions of synapses are associated with a number of neurological and psychiatric disorders, including Alzheimer's, schizophrenia, and depression.

Synaptic weight see **Weight.**

Test accuracy *Accuracy.* A metric used in machine learning to measure the performance of a model. It is defined as the ratio between the correctly predicted or classified objects and the total number of objects in the dataset. For example, if a model that has been trained to classify images correctly classifies 90 out of 100 images, then the test accuracy of the model is 90%.

Test dataset Part of a dataset that is used to test an already trained neural network or model. Typically 20% of the total dataset. See also **Dataset splitting.**

Top-down From hierarchically higher to lower processing levels.

Training dataset Part of a dataset that is used for training a neural network or other machine learning model. Typically 80% of the total dataset. See also **Dataset splitting.**

Transfer learning A technique of machine learning where a model is first trained on a large dataset and then refined on a smaller dataset for a specific task *(Fine Tuning).* The idea of transfer learning is that the knowledge gained in solving one problem can be transferred to another, related problem, thereby reducing the amount of data and time required for training a new model.

Transformer Neural network architecture that is particularly suitable for processing natural language, e.g., translation and text generation. Unlike recurrent neural networks (RNN) and convolutional networks (CNN), the Transformer uses a so-called **attention mechanism,** which allows the model to selectively focus on different parts of the input sequence to make predictions. A Transformer consists of an encoder and a decoder, both of which are made up of multiple neural network and attention layers. The encoder processes the input sequence and generates internal representations from it, which the decoder uses to generate the output sequence. Transformers have several advantages over traditional neural networks, for example, the ability to process input sequences in parallel, handle sequences of variable length, or capture long-range dependencies in sequences without suffering from the problem of vanishing/exploding gradients. See also **ChatGPT, GPT-3, GPT-4** and **Large Language Model.**

Tri-Level Hypothesis Theoretical framework proposed by David Marr in the field of cognitive science and artificial intelligence. According to this, every natural or artificial system that performs a cognitive task can be described on three levels of analysis. The **computational level** describes the problem to be solved, the goal of the system (brain or AI), and the constraints imposed by the environment. It specifies what information needs to be processed, what output needs to be generated, and why the system needs to solve the problem. The **algorithmic level** describes the rules and procedures, i.e., the algorithm, that the system must follow to solve the problem specified at the computational level. It specifies how the input data is transformed into output data and how the system processes information. The **implementation level** describes the physical implementation of the system, e.g., the hardware and software used to build the AI. It specifies the details of how the algorithmic level is implemented, including the data structures used, programming languages, and computational resources. In neurobiology, the anatomical and physiological details of the nervous system are described at this level. According to Marr, it is necessary to understand a system at all three levels in order to fully grasp its behavior and possibly develop more efficient systems.

Turing Test A method proposed by Alan Turing and originally called the *Imitation Game* by him to test a machine's ability for intelligent behavior. In the simplest variant, one or more human examiners communicate in natural language text-based, i.e., in the form of a chat, both with a human (as a control) and with the machine to be tested, without knowing who is who. If the majority of examiners cannot reliably distinguish between human and machine, the Turing Test is considered passed for the machine. This variant of the Turing Test was passed in 2022 by the AI system **ChatGPT** as the texts and responses it generated are indistinguishable from those produced by humans. In principle, there are other, more difficult variants of the Turing Test. For example, the dialogue can take place not as a pure chat, but as an actual conversation in spoken language, similar to a telephone conversation. In this case, the machine would also have to be able to correctly interpret and imitate linguistic features such as emphasis and sentence melody. Finally, the AI could also be integrated into a humanoid robot. In this variant, which is also referred to as the Embodied Turing Test, the entire spectrum of human behavior including facial expressions, gestures, and all motor skills would have to be imitated, and the Turing Test would ultimately aim to convince other people that the machine is also a human. In the final escalation, the machine would not only have to convince others, but also be convinced itself that it is a human, and would not be allowed to know or suspect anything about its true nature. Such scenarios have been picked up in films like *Imposter* or the series *Westworld* and the consequences have been thought through to the end. The Turing Test has been much discussed in the field of Artificial Intelligence. Some critics argue that it sets an unrealistic standard for intelligence, as there are many tasks that humans can perform, but machines cannot, and vice versa. Others argue that passing the Turing Test does not necessarily mean that a machine is truly intelligent, but rather that it is good at imitating human behavior. One of the most famous counterarguments to the Turing Test is John Searle's thought experiment on the **Chinese Room.** Despite this criticism, the Turing Test remains an important benchmark for research in the field of Artificial Intelligence, and many researchers continue to work on the development of machines that can pass the test.

Unsupervised Learning Type of machine learning in which a model learns to extract useful patterns and structures from data without the need for labeled data. This type of learning typically involves tasks such as clustering, where similar data points are grouped, and dimensionality reduction, where high-dimensional data is represented in a low-dimensional space while preserving important information. Examples of unsupervised learning algorithms include K-Means clustering, hierarchical clustering, and Principal Component Analysis (PCA).

Vertical Refers to the processing direction between elements on different hierarchy levels as opposed to horizontal processing between elements of the same hierarchy level. See also **Top-down** and **Bottom-up.**

Word embedding A method of machine learning and natural language processing in which each word is assigned a **word vector.**

Weight Also Synaptic weight. Strength of the connection between two artificial or natural neurons. The weight can in principle be any real number, with the magnitude representing the size of the effect on the successor neuron and the sign representing the quality of the effect (excitation or inhibition).

Weight matrix Matrix whose entries contain the weights between a set of neurons. The matrix can contain all pairwise weights of all neurons of a neural network, including all self-connections (autapses). In this case, it is referred to as a *complete weight matrix.* However, a weight matrix can also only contain the forward-directed weights between the neurons of two successive layers of a network. The columns or rows of the weight matrix correspond to the input or output weight vectors of the neurons.

Weight vector Vector whose entries contain all the weights of a neuron. A distinction is made between input weight vectors, which contain the weights of a neuron with which its input is weighted, and output weight vectors, which contain the weights to the successor neurons of a neuron.

Word vector Representation of the meaning of a word as a vector. The more different the meaning of two words is, the more different are the corresponding word vectors. If you interpret the word vectors as points in a semantic space, the distance between the points corresponds to the similarity or dissimilarity of the underlying words. The smaller the distance, the more similar the meaning. Synonyms, i.e., words with the same meaning, are mapped to the same word vector or point, so they have a distance of zero. Similarly, there are also **sentence vectors,** which represent the meaning of an entire sentence.

XOR Problem A classic problem in the field of Artificial Intelligence and Machine Learning that illustrates the limitations of certain types of models. XOR stands for eXclusive OR, a binary operation in which the output is true (or 1) only when the number of true inputs is odd. With two binary inputs, XOR is true only when exactly one of the inputs is true. The XOR problem refers to the challenge of correctly classifying these four situations using a linear classifier, e.g., a single-layer perceptron. The problem is that the XOR function is not linearly separable, i.e., there is no straight line (in 2D space) that can separate the inputs that yield a 1 from those that yield a 0. This illustrates the inability of linear classifiers to handle certain types of patterns. Multi-layer neural networks, however, can solve the XOR problem by creating non-linear decision boundaries. This is typically achieved by introducing hidden layers and non-linear activation functions into the network.

Zero-Shot Learning A form of machine learning where a model is trained to recognize objects or categories it has never seen before. It can classify new input patterns even when no labeled data for the relevant class were available during training. Unlike supervised learning, where a model is trained with a specific set of labeled data examples, zero-shot learning is based on the transfer of knowledge from related or similar classes that were seen during training. This is achieved by using semantic representations such as word vectors that capture the meaning and relationships between different classes. For example, if a model has been trained to recognize images of animals and has never seen an image of a zebra, it can still classify it as an animal because it has learned the relationships between different animal species. Zero-shot learning enables more efficient and flexible training of models and generalization to new and unknown categories.

Printed in the United States
by Baker & Taylor Publisher Services

Printed in the United States
by Baker & Taylor Publisher Services